Temporal Bone CT and MRI Anatomy

Jan Kopřiva • Jan Žižka

Temporal Bone CT and MRI Anatomy

A Guide to 3D Volumetric Acquisitions

Springer

Jan Kopřiva, MD
Department of Radiology
University Hospital Hradec Králové
Hradec Králové
Czech Republic

Jan Žižka, MD, PhD
Department of Radiology
University Hospital Hradec Králové
Charles University in Prague
Faculty of Medicine in Hradec Králové
Hradec Králové
Czech Republic

ISBN 978-3-319-08241-7 ISBN 978-3-319-08242-4 (eBook)
DOI 10.1007/978-3-319-08242-4
Springer Cham Heidelberg New York Dordrecht London

Library of Congress Control Number: 2014954745

Printed on acid-free paper

Springer is part of Springer Science+Business Media (www.springer.com)

Foreword

Understanding the anatomy of the temporal bone has always been, is and will be a difficult task for doctors of various specialities: radiology, otorhinolaryngology, neurosurgery or anatomy.

Temporal Bone CT and MRI Anatomy by Jan Kopřiva and Jan Žižka is a comprehensive atlas of radiological anatomy of the temporal bone. All important anatomical structures of the temporal bone are didactically and precisely labeled in three standard planes of view.

I personally recommend a thorough survey of the black-and-white images within this CT and MRI atlas to every doctor who is interested in the given field. The presented knowledge forms a mere basic cornerstone in gradually acquired experience in clinical CT and MRI of the temporal bone.

This atlas may aid in promoting consultations and discussions between radiologists and otologists, as only two pairs of eyes and two individual brains lead to true understanding of the complicated beauty of the temporal bone where the black-and-white radiologist's view turns into a colourful clinical picture of the correctly diagnosed disorder.

Hradec Králové, Czech Republic — Viktor Chrobok, MD, PhD

Preface

A few years ago, I was asked by Professor Viktor Chrobok, the head of the Department of Head and Neck Surgery, University Hospital Hradec Králové, to prepare an educational material on temporal bone anatomy. At that time, I could hardly anticipate how much the temporal bone will influence my life. The temporal bone – feared by so many students of anatomy as well as residents of radiology or head and neck surgery; their companion during sleepless nights spent over the anatomy textbooks; the most complex and difficult to understand bone of the human body.

There is no doubt that knowledge of anatomy is the fundamental prerequisite to proper diagnosis making. I was happy enough to gain my applied anatomy skills during my professional 2-year fellowship at the Department of Anatomy at the Charles University Medical School in Hradec Králové. I took the opportunity to deepen my knowledge of the spatial arrangement of the middle ear during the temporal bone preparative courses led by Prof. Viktor Chrobok. Later on when I (already as a radiology resident) focused my professional interest on the details of the temporal bone CT anatomy, I simply did not find a textbook which would cover this topic in its full complexity. And this, in fact, turned out to be the key issue in the inception of this atlas.

Hradec Králové, Czech Republic — Jan Kopřiva, MD

Scope

It is an indisputable fact that knowledge of anatomy is one of the cornerstones of clinical medicine. Nowadays, with the advancement of microsurgical techniques and state-of-the-art imaging modalities, this applies to surgical specialities as well as to radiology more than ever. Familiarity with the key anatomical structures and their basic topographic relationships is necessary for the effective use of this atlas, which does not aspire to be a fundamental source of information for medical students. That is, among other things, why we have not included currently so popular and visually impressive 3D reconstructions, mostly suitable for demonstration of the surface anatomy of the temporal bone: this topic is extensively covered by respective textbooks and atlases.

Our major goal is to provide the reader with the detailed temporal bone imaging anatomy in the same way as it is routinely displayed on contemporary DICOM viewing consoles, i.e. as a series of consecutive sub-millimetre sections in standardized orientation.

Acknowledgements

Our sincere thanks belong to Prof. Viktor Chrobok, MD, PhD, who stood at the very roots of this atlas, and to our colleagues Zdeněk Bělobrádek, MD, PhD, and Jiří Vaňásek, MD, for their technical support. Our special thanks belong to Olga Čermáková for her diligent assistance with the graphic layout of the book.

Abbreviations

AA	anterior ampulla
AAA	aditus ad antrum
	entrance to the mastoid antrum
ACN	accessory nerve
	n.XI
ACP	air cell of the pyramid
AE	anterior epitympanum
AICA	anterior inferior cerebellar artery
AN	abducens nerve
	n.VI
AOA	anterior osseous ampulla
ASC	anterior semicircular canal
	also called superior semicircular canal
BA	basilar artery
BI	body of incus
	also called corpus incudis
BSL	bony spiral lamina
	also called lamina spiralis ossea
CA	cochlear area
	entry of cochlear nerve, portion of n.VIII
CAC	carotid canal
	contains internal carotid artery
CAQ	cochlear aqueduct
	also called perilymphatic duct or aquaeductus cochleae
CAT	cochlea – apical turn
CBT	cochlea – basilar turn
CCR	common crus
CE	cerebellum
CJS	caroticojugular spine

CMT cochlea – middle turn
CN cochlear nerve
  portion of n.VIII
COC common osseous crus
COCA condylar canal
  communication between extra- and intracranial veins
CP cochleariform process
CPC cerebellopontine cistern
CTAT crista transversa – anterior epitympanum
E epitympanum
  also called attic
EAC external auditory canal
EACA external aperture of the cochlear aqueduct
EOVA external opening of the vestibular aqueduct
ET Eustachian tube
FAC falx cerebelli
FC flocculus
FL foramen lacerum
FN facial nerve
  n.VII
FNA facial nerve area
  entry of the facial nerve
FNCLS facial nerve canal – labyrinthine segment
  contains labyrinthine segment of the facial nerve
FNCMS facial nerve canal – mastoid segment
  contains mastoid segment of the facial nerve
FNCTS facial nerve canal – tympanic segment
  contains tympanic segment of the facial nerve
FNES facial nerve – extracranial segment
FNFG facial nerve first genu
FNSG facial nerve second genu
FO foramen ovale
  contains mandibular nerve
FP foramen petrosum
  contains lesser petrosal nerve
FR facial recess
FS foramen spinosum
  contains middle meningeal artery
FV fourth ventricle
GF geniculate fossa
  contains geniculate ganglion
GIPS groove of the inferior petrosal sinus
  contains inferior petrosal sinus
GN glossopharyngeal nerve
  n.IX

GSS groove of the sigmoid sinus
GWS greater wing of sphenoid
HC hypoglossal canal
  contains twelfth cranial nerve and venous plexus
HN hypoglossal nerve
  n.XII
HT hypotympanum
I incus
IAC internal auditory canal
IACF internal auditory canal – fundus
IAM internal auditory canal – meatus
ILP incus – long process
IMJ incudomallear joint
ISJ incudostapedial joint
ISP incus – short process
  also called short crus or limb
IVA inferior vestibular area
  entry of the saccular nerve
JB jugular bulb
JF jugular foramen
JFPN jugular foramen – pars nervosa
JFPV jugular foramen – pars vascularis
JS jugular spine
KS Koerner's septum
LA lateral ampulla
LOA lateral osseous ampulla
LP lenticular process
LPE lateral posterior epitympanum
LSC lateral semicircular canal
  also called horizontal or external semicircular canal
MA mastoid antrum
MAC mastoid air cells
MACA mastoid canaliculus
  contains auricular branch of the vagus nerve
MAN mastoid notch
MAP malleus – anterior process
MCBS mastoid cells bony septa
MCF middle cranial fossa
MCP middle cerebellar peduncule
MET mesotympanum
MF mandibular fossa
MH malleus – head
MHA malleus – handle
MHE mandible – head
MLP malleus – lateral process

MN malleus – neck
MNE mandible – neck
MO medulla oblongata
MOD modiolus
MP mastoid process
MPE medial posterior epitympanum
OB occipital bone
OMS occipitomastoid suture
OW oval window
OWN oval window niche
P promontory
PC petromastoid canal
    contains subarcuate artery and vein
PCF posterior cranial fossa
PCP posterior clinoid process
PE petrous part of the temporal bone
PLSC prominence of the lateral semicircular canal
PMC premedullary cistern
    also called perimedullary cistern
PMSTB posteromedial surface of the temporal bone
PNS pons
PO ponticulus
POA posterior osseous ampulla
POF petrooccipital fissure
POT posterior tympanum
PPC prepontine cistern
PS Prussak's space
PSC posterior semicircular canal
PT protympanum
PTF petrotympanic fissure
    also called glaserian, an exit route for the chorda tympani
PYE pyramidal eminence
RW round window
RWN round window niche
SAC stapes – anterior crus
SAF subarcuate fossa
    entry to the petromastoid canal
SC simple crus
SCT scutum
SIC singular canal
    entry of the posterior ampullar nerve
SIT sinus tympani
SM stapedius muscle
SMF stylomastoid foramen

SN saccular nerve
    portion of n.VIII, also called inferior branch of the vestibular nerve
SOC simple osseous crus
SP styloid process
SPC stapes – posterior crus
SPF sphenopetrosal fissure
SPS sphenoidal sinus
SQP squamous part of the temporal bone
    also called squama temporalis
SQT squama temporalis
    also called squamous part of the temporal bone
SSS sphenosquamosal suture
STH stapes – head
SU subiculum
SVA superior vestibular area
    entry of the utriculo-ampullar nerve
TC tympanic canaliculus
    contains tympanic nerve and inferior tympanic artery
TCR transverse crest
    also called crista transversa
TL temporal lobe
TM tympanic membrane
TMF tympanomastoid fissure
TN trigeminal nerve
    n.V
TSF tympanosquamous fissure
TTM tensor tympani muscle
TTMT tensor tympani muscle – tendon
UAN utriculo-ampullar nerve
    portion of n.VIII, also called superior branch of the vestibular nerve
V vestibule
VA vertebral artery
VAQ vestibular aqueduct
VCN vestibulocochlear nerve
    n.VIII
VER vestibule – elliptical recess
VN vagus nerve
    n.X
VSR vestibule – spherical recess
ZP zygomatic process

Contents

Introduction

Dear Reader,

There were two major reasons why this book came into existence, eventually: extensive progress and spread of multidetector CT technology (MDCT) and the absence of similarly focused publications on the market. The major aim of this book is to provide the reader with a comprehensive, systematic and topical view of the anatomy of the temporal bone from the perspective of advanced imaging methods, namely MDCT and MRI. We have intended to provide the reader with complete image datasets comprising consecutive high-spatial-resolution CT and MRI slices of sub-millimetre thickness, presented in the same way as they emanate from contemporary MDCT and MRI scanners, with all anatomical structures labelled. This is why we have not included any histological sections, developmental variants or pathologic conditions which are discussed elsewhere.

The book addresses primarily radiologists, although we hope that it may offer valuable information also to head and neck surgeons, neurosurgeons as well as anatomists.

The Authors

Part I
Imaging Issues

Chapter 1
Image Quality and Radiation Dose Aspects

As the anatomy of the temporal bone comprises of very small structures, high spatial resolution imaging techniques are a prerequisite for adequate diagnostic results. For several decades, the effort has been aimed at maximizing the spatial resolution as well as minimizing the image noise.

1.1 MDCT

In case of high-resolution CT (HRCT) of the pyramids, the scanning protocols were traditionally associated with relatively high radiation doses (e.g. effective tube current–time product of 320 mAs and peak tube voltage of 120 kVp), yielding up to 75 mGy of volume CT dose index ($CTDI_{vol}$). These scanning parameters were comparable or even exceeding those of routine CT head scanning protocols. Moreover, protocols based on 140 kVp tube voltage settings commonly exceeded $CTDI_{vol}$ values of 90 mGy, thus further increasing the radiation load near/in the area of the eye lens which is the most endangered organ in terms of radiation damage in adult head CT scans. It should also be noted that the vast majority of current helical MDCT head scanning protocols do not employ or even do not allow gantry tilt. The eye lens is therefore regularly exposed to the primary beam in both head and temporal bone MDCT scanning procedures. Furthermore, on conventional as well as multidetector CT scanners equipped with less than 16 rows of detectors, head CT scans were routinely acquired separately from HRCT studies of the temporal bones, thus virtually doubling the radiation dose to the eye lens.

According to the latest statement on tissue reactions issued by the International Commission on Radiological Protection (ICRP), the threshold in absorbed dose for the eye lens was lowered from 2.0 to 0.5 Gy in order to minimize the risk of radiation-induced cataract development.

Whereas the radiation protection measures and ALARA ("as low as reasonably achievable") principles urge us on minimizing the radiation dose, the abovementioned

J. Kopřiva, J. Žižka, *Temporal Bone CT and MRI Anatomy: A Guide to 3D Volumetric Acquisitions*, DOI 10.1007/978-3-319-08242-4_1

scanning strategies and continuously increasing numbers of CT procedures worldwide act in contradiction.

Fortunately, the state-of-the-art MDCT scanners offer numerous options on how to effectively reduce the radiation dose without sacrificing spatial resolution and diagnostic performance. First, their detector arrays of submillimetre (0.5–0.6 mm) width allow for acquiring high spatial resolution 3D volumetric datasets: the submillimetre collimation necessary for HRCT is thus automatically available in every MDCT acquisition. Second, increased quantum efficiency of current detectors allows for lowering kVp and mAs settings in all acquisitions without significant increase in image noise. Third, novel techniques of iterative CT image reconstruction benefit from inherent low image noise levels and have already proved their abilities to decrease radiation exposure by tens of percent when compared to standard filtered back projection which dominated CT technology for more than three decades.

Current MDCT systems are therefore routinely capable of reconstructing HRCT images of the skull base and pyramids with isotropic submillimetre resolution from every routine head CT scan, with markedly reduced dose compared to conventional ($CTDI_{vol} \approx 60$ mGy) standards and without additional radiation burden, as both thin collimation raw data and high spatial resolution reconstructive algorithm ("sharp" or "bone" kernel) are readily available.

We aimed to demonstrate that with the adoption of adequate precautions focused on radiation dose reduction, it is feasible to obtain 3D volumetric datasets offering both high spatial resolution and sufficient diagnostic image quality. We would like to encourage the users of state-of-the-art MDCT scanners to acquire image data for both cerebral and temporal bone studies from a single helical acquisition, thus avoiding the need for any additional scanning of the temporal bones in routine head MDCT scans and minimizing the radiation dose impact.

The protocol we routinely use in our institution for non-emergency helical head MDCT scans combines iterative reconstruction algorithm with automatic tube current modulation and carefully reduced reference mAs settings in order to reduce both the effective and particularly the organ dose to the eye lens. When compared to the European Commission Quality Criteria for MDCT which set $CTDI_{vol}$ of 60 mGy as a reference standard for head scans, our protocol yields a 53 % dose reduction when compared to the reference standard (28.2 mGy versus 60 mGy).

All HRCT scans showed in this book were acquired in a helical mode, with no gantry tilt, on a single source MDCT scanner with the following parameters: collimation $2 \times 64 \times 0.6$ mm, tube voltage of 120 kV, pitch factor of 0.55, rotation time of 1 s, and quality reference mAs values set to 240 mAs. The iterative reconstruction kernel was set to J70s for HRCT of the pyramids (or alternatively to J30s for cerebral scans, not shown). Slice thickness of all HRCT scans shown in the book is 0.6 mm with the reconstruction interval of 0.4 mm. The average effective dose of this protocol is 1.18 mSv. The average organ dose to the eye lens yields 31.9 mGy.

It is an unfavourable fact that the average medical radiation dose to the population of the most developed countries showed an increase of several hundred percent in a single generation. The authors hereby want to demonstrate that a proper use of dose-reducing strategies is capable of significant radiation dose reduction, namely, in terms of organ dose delivered to the eye lens, which is more radiosensitive than previously thought. The synergic effect of the utilized advanced techniques allowed us to routinely obtain helical head MDCT scans with only 47 % of the radiation dose set as a reference standard for head scans, with the additional bonus of obtaining high-quality HRCT of the pyramids in every routine head MDCT examination "for free".

1.2 MRI

MRI offers a complementary information to HRCT, with excellent visualization of cranial nerves and inner ear fluid containing structures, thus completing the "skeletal" type of information obtained from MDCT with the "neural" component available from MRI.

Principally, there are two types of MRI sequences suitable for high spatial resolution imaging of the cranial nerves and inner ear anatomy: either balanced steady-state 3D gradient echo (CISS, FIESTA, bFFE) or 3D fast/turbo spin echo (SPACE, VISTA, CUBE, VFA) with artificially restored longitudinal magnetization which helps keep the acquisition time within acceptable ranges (Restore, DRIVE, FRFSE-XL, FIESTA). Both approaches yield similar results with comparable acquisition times; nonetheless, 3D FSE/TSE techniques are less prone to inherent susceptibility artefacts which might degrade the image quality in the proximity of bony structures.

It is feasible to obtain the submillimetre spatial resolution fully comparable with that of MDCT in all three planes using current MRI systems. MRI scans presented in this book were acquired on a clinical 1.5 T MRI system with the following parameters: TR = 1,400 ms, effective TE = 259 ms, turbo factor (ETL) = 85, PAT factor = 2, voxel size of 0.6 × 0.6 × 0.6 mm, and acquisition time of 6:00 mins.

Part II
The Atlas

Structure

Despite the fact that interactive online anatomical applications are widely popular at present, we have chosen a printed issue of this topic. Not only just because books last for ages. The interactivity is substituted here by means of complexity of the data: complete sets of source axial MDCT and MR images are accompanied by reformatted CT sections in coronal and Stenverse planes (i.e. in the plane perpendicular to the course of the superior semicircular canal). The anatomy is listed here step-by-stepwise or, if you wish, slice-by-slice, resembling the way you scroll through the images on a diagnostic workstation. Therefore, the motto of this book might read "what I can see on the screen is what I get in the book".

The image atlas is divided into four chapters: (1) axial HRCT sections listed in craniocaudal direction, (2) coronal HRCT sections in anteroposterior direction, (3) Stenverse HRCT sections in lateromedial direction, and (4) axial MRI sections in craniocaudal direction. For readers' convenience, all anatomical structures are labelled in each image. The total number of anatomical structures identified in this atlas exceeds 150.

Histological sections have been intentionally omitted in this book because this topic had been extensively covered by other literary sources. Similarly, pathologic conditions of the temporal bone have not been included in this issue as these are extensively discussed in other sources like *Imaging of the Temporal Bone* by Swartz and Loevner.

Chapter 2
HRCT Axial Reformat

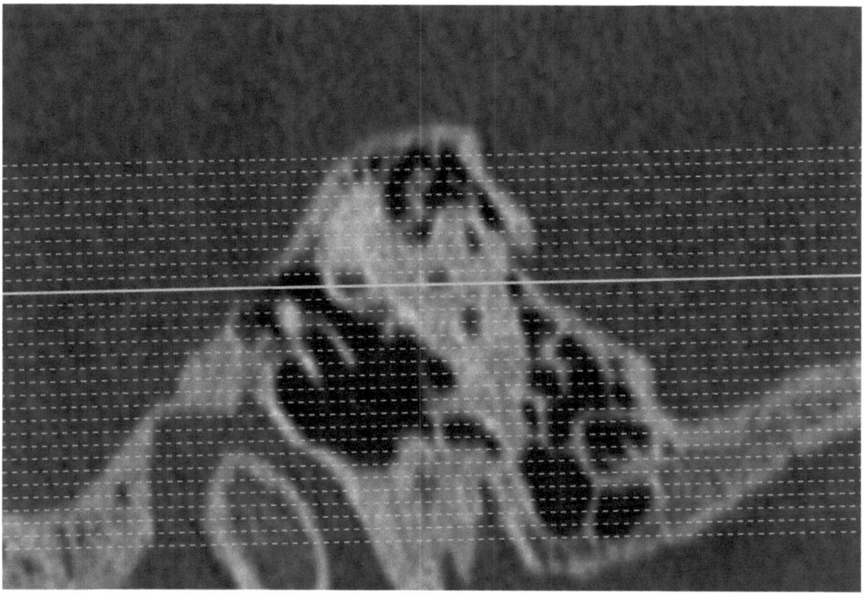

A00: HRCT axial reformat
Axial plane:

Using an MPR (multi-planar reformat) 3D viewer, anterior and posterior aspects of the lateral semicircular canal are identified. Their connecting line defines the axial plane.

J. Kopřiva, J. Žižka, *Temporal Bone CT and MRI Anatomy: A Guide to 3D Volumetric Acquisitions*, DOI 10.1007/978-3-319-08242-4_2

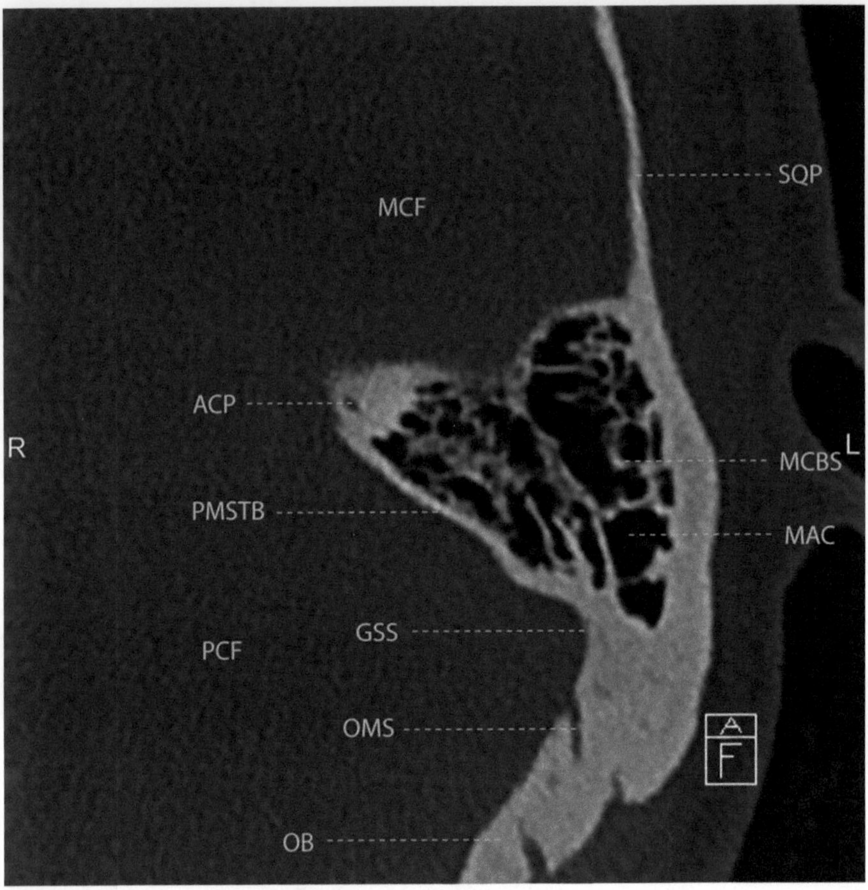

A01: HRCT axial reformat

ACP	air cell of the pyramid
GSS	groove of the sigmoid sinus
MAC	mastoid air cells
MCBS	mastoid cells bony septa
MCF	middle cranial fossa
OB	occipital bone
OMS	occipitomastoid suture
PCF	posterior cranial fossa
PMSTB	posteromedial surface of the temporal bone
SQP	squamous part of the temporal bone

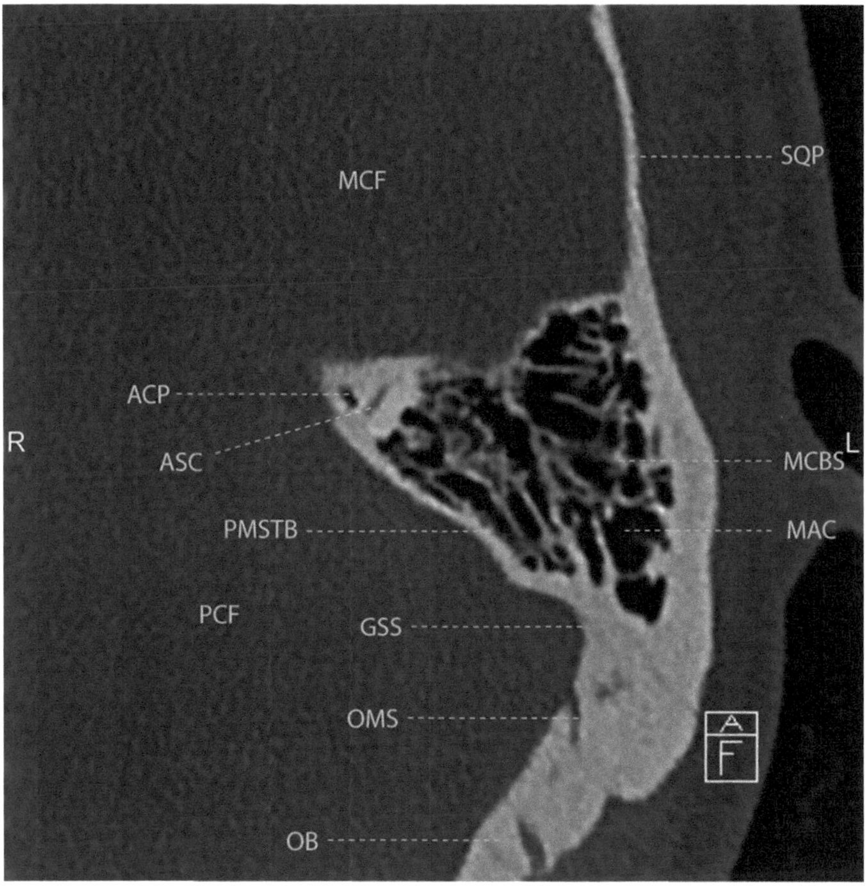

A02: HRCT axial reformat

ACP	air cell of the pyramid	OMS	occipitomastoid suture
ASC	anterior semicircular canal	PCF	posterior cranial fossa
GSS	groove of the sigmoid sinus	PMSTB	posteromedial surface of the temporal bone
MAC	mastoid air cells	SQP	squamous part of the temporal bone
MCBS	mastoid cells bony septa		
MCF	middle cranial fossa		
OB	occipital bone		

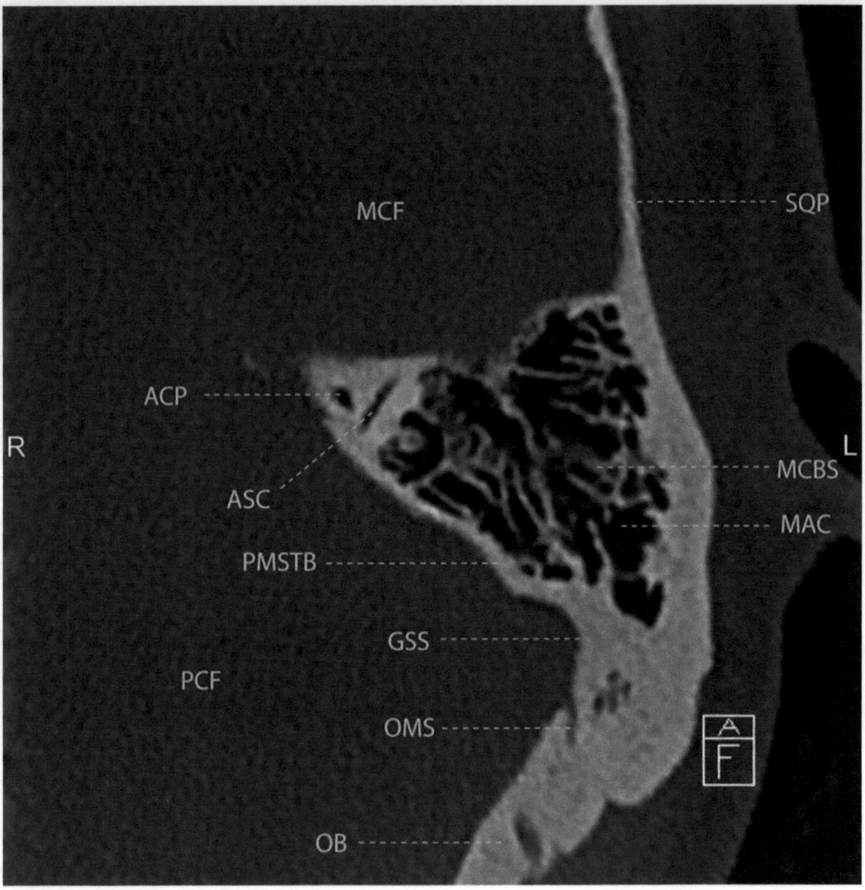

A03: HRCT axial reformat

ACP	air cell of the pyramid	OMS	occipitomastoid suture
ASC	anterior semicircular canal	PCF	posterior cranial fossa
GSS	groove of the sigmoid sinus	PMSTB	posteromedial surface of the temporal bone
MAC	mastoid air cells	SQP	squamous part of the temporal bone
MCBS	mastoid cells bony septa		
MCF	middle cranial fossa		
OB	occipital bone		

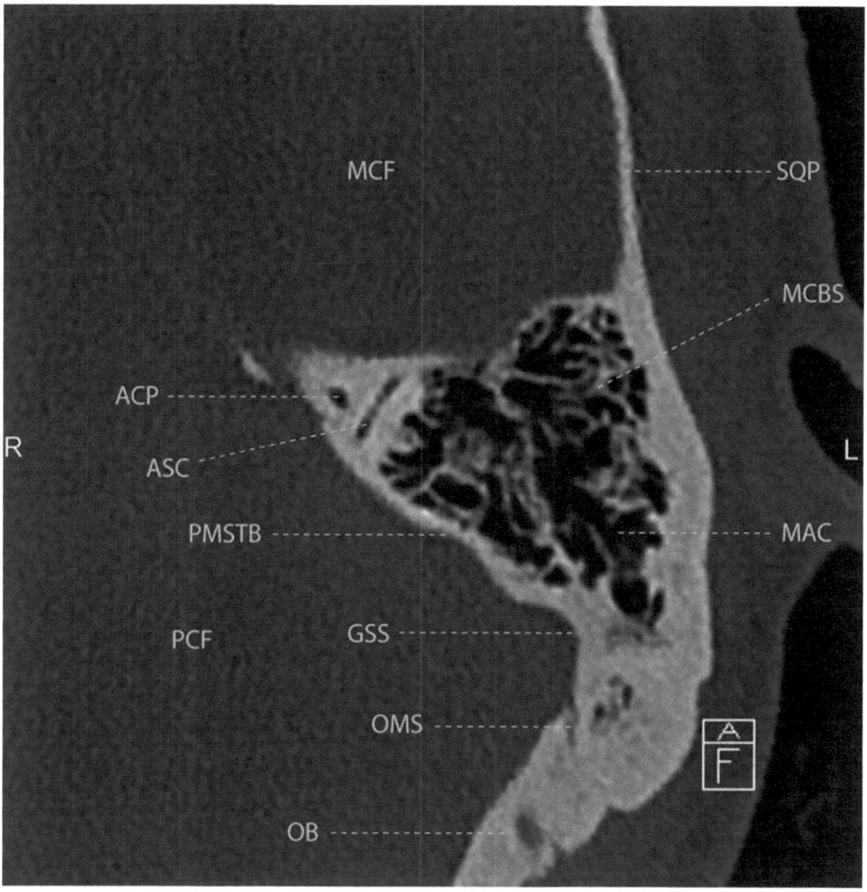

A04: HRCT axial reformat

ACP	air cell of the pyramid	OMS	occipitomastoid suture
ASC	anterior semicircular canal	PCF	posterior cranial fossa
GSS	groove of the sigmoid sinus	PMSTB	posteromedial surface of the temporal bone
MAC	mastoid air cells		
MCBS	mastoid cells bony septa	SQP	squamous part of the temporal bone
MCF	middle cranial fossa		
OB	occipital bone		

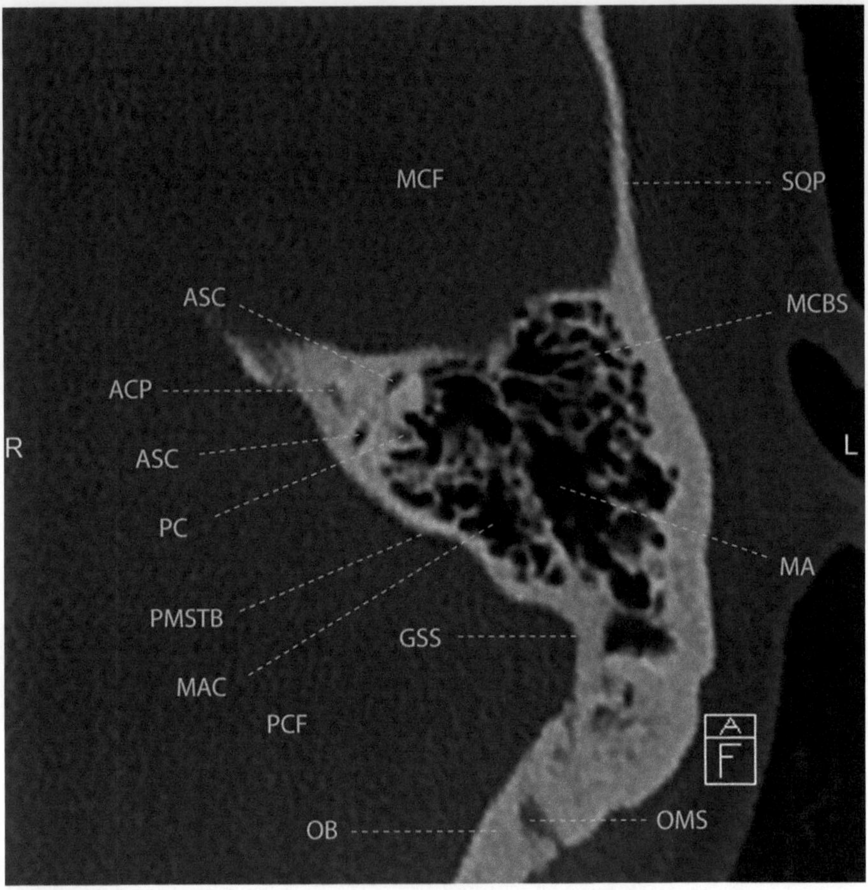

A05: HRCT axial reformat

ACP	air cell of the pyramid
ASC	anterior semicircular canal
GSS	groove of the sigmoid sinus
MA	mastoid antrum
MAC	mastoid air cells
MCBS	mastoid cells bony septa
MCF	middle cranial fossa
OB	occipital bone
OMS	occipitomastoid suture
PC	petromastoid canal
PCF	posterior cranial fossa
PMSTB	posteromedial surface of the temporal bone
SQP	squamous part of the temporal bone

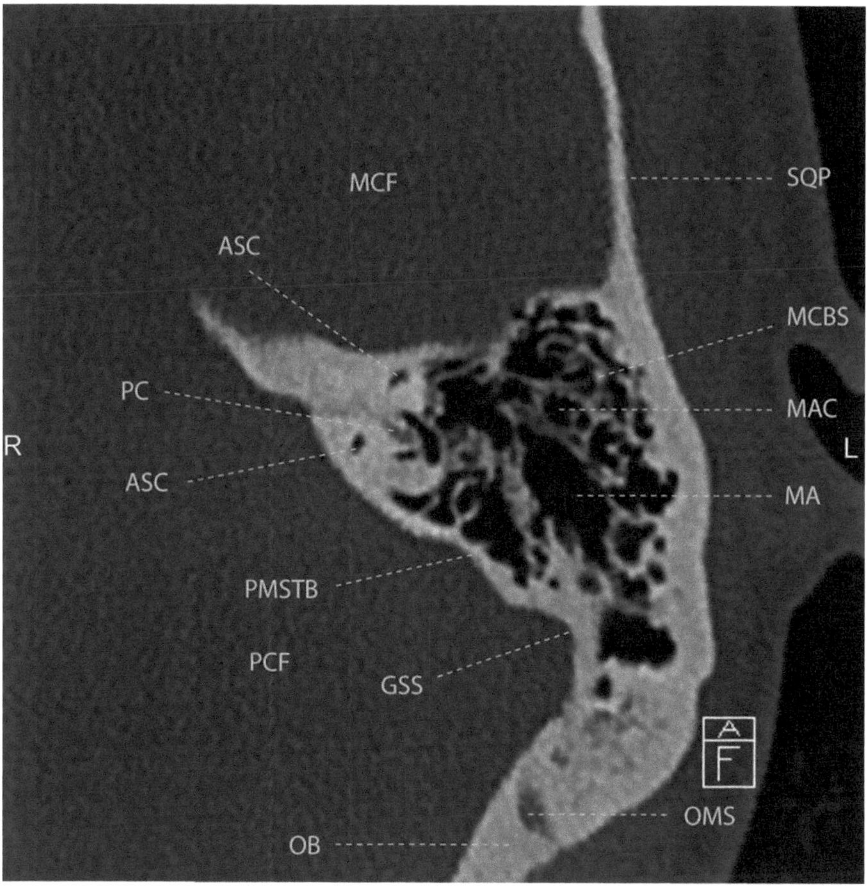

A06: HRCT axial reformat

ASC	anterior semicircular canal	OMS	occipitomastoid suture
GSS	groove of the sigmoid sinus	PC	petromastoid canal
MA	mastoid antrum	PCF	posterior cranial fossa
MAC	mastoid air cells	PMSTB	posteromedial surface of the temporal bone
MCBS	mastoid cells bony septa	SQP	squamous part of the temporal bone
MCF	middle cranial fossa		
OB	occipital bone		

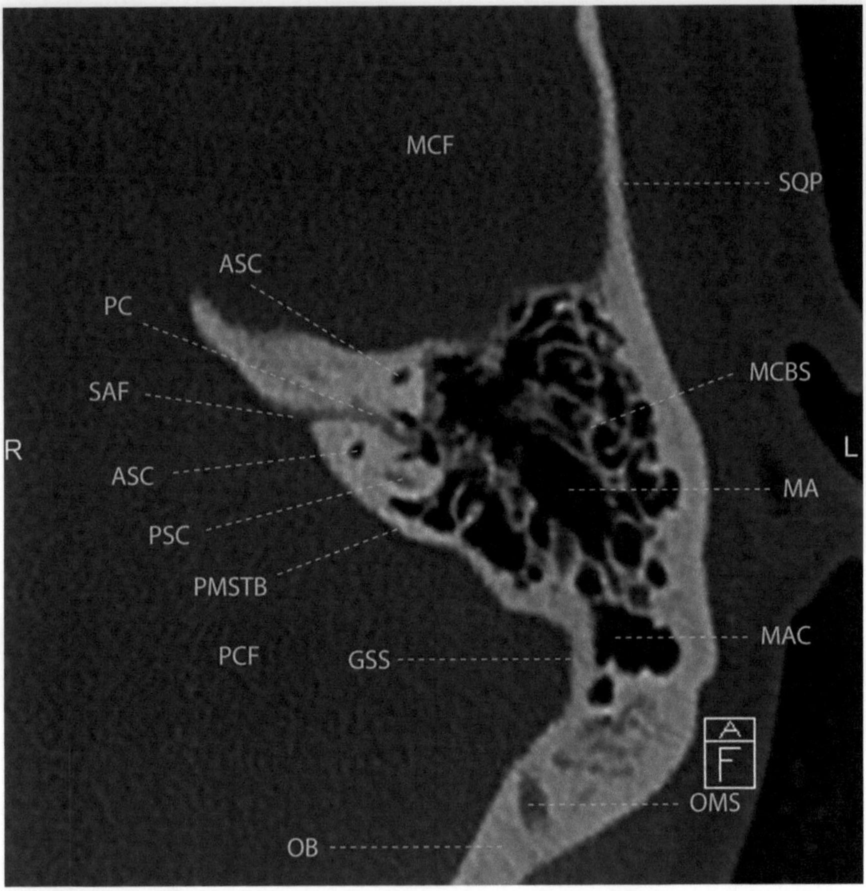

A07: HRCT axial reformat

ASC	anterior semicircular canal
GSS	groove of the sigmoid sinus
MA	mastoid antrum
MAC	mastoid air cells
MCBS	mastoid cells bony septa
MCF	middle cranial fossa
OB	occipital bone
OMS	occipitomastoid suture
PC	petromastoid canal
PCF	posterior cranial fossa
PMSTB	posteromedial surface of the temporal bone
PSC	posterior semicircular canal
SAF	subarcuate fossa
SQP	squamous part of the temporal bone

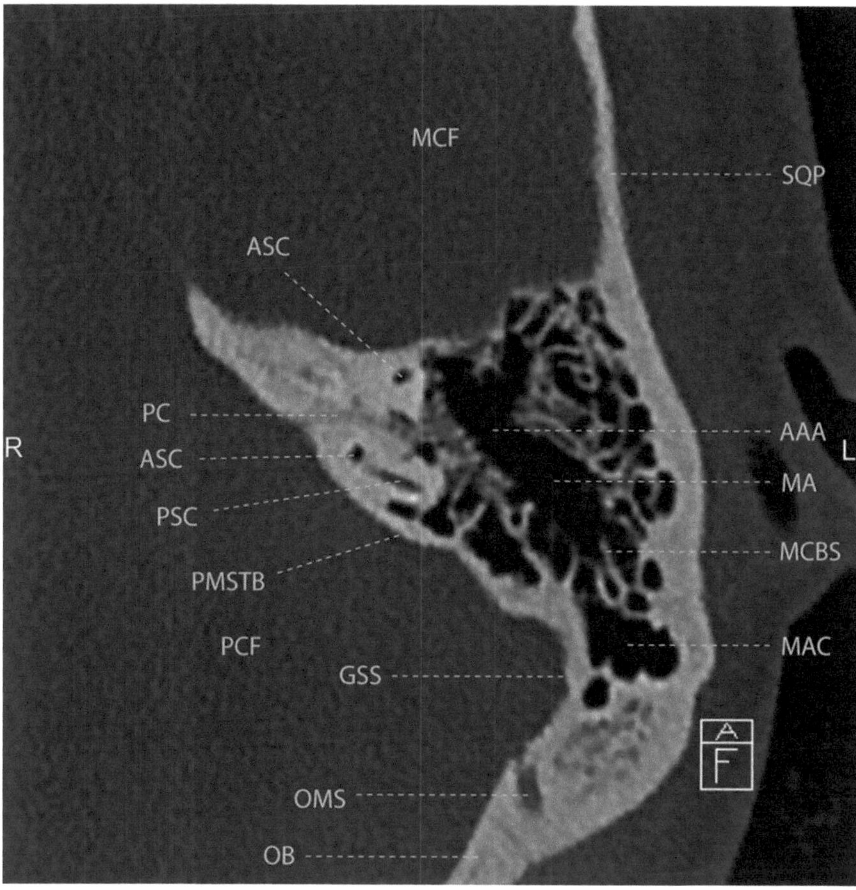

A08: HRCT axial reformat

AAA	aditus ad antrum	OMS	occipitomastoid suture
ASC	anterior semicircular canal	PC	petromastoid canal
GSS	groove of the sigmoid sinus	PCF	posterior cranial fossa
MA	mastoid antrum	PMSTB	posteromedial surface of the temporal bone
MAC	mastoid air cells		
MCBS	mastoid cells bony septa	PSC	posterior semicircular canal
MCF	middle cranial fossa	SQP	squamous part of the temporal bone
OB	occipital bone		

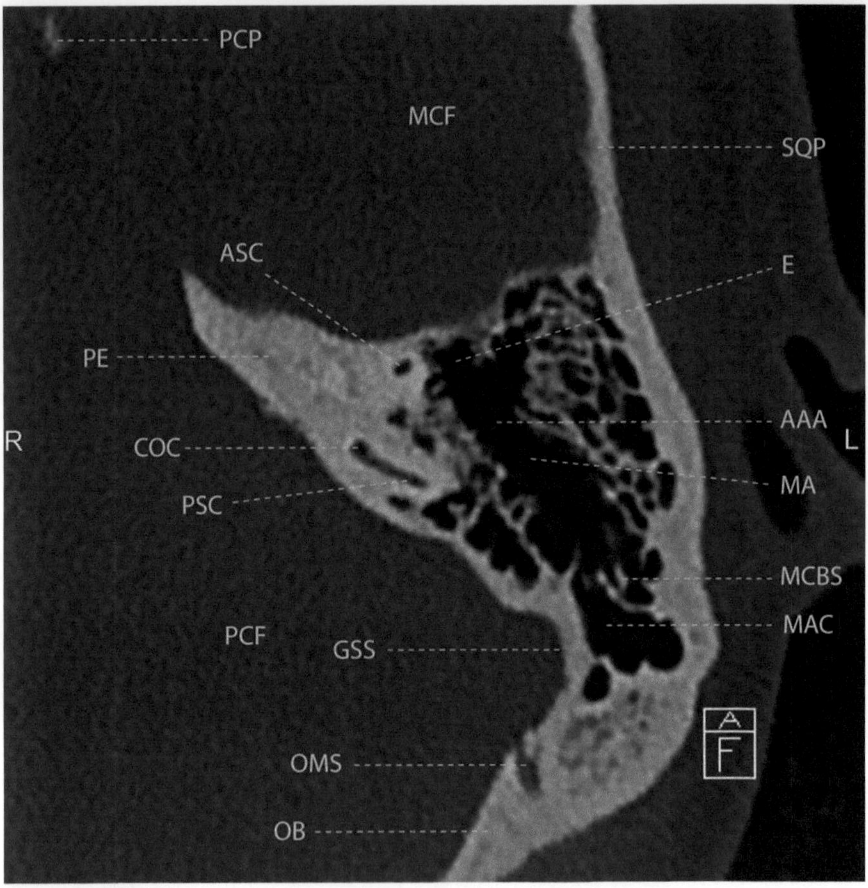

A09: HRCT axial reformat

AAA	aditus ad antrum	OB	occipital bone
ASC	anterior semicircular canal	OMS	occipitomastoid suture
COC	common osseous crus	PCF	posterior cranial fossa
E	epitympanum	PCP	posterior clinoid process
GSS	groove of the sigmoid sinus	PE	petrous part of the temporal bone
MA	mastoid antrum	PSC	posterior semicircular canal
MAC	mastoid air cells	SQP	squamous part of the temporal bone
MCBS	mastoid cells bony septa		
MCF	middle cranial fossa		

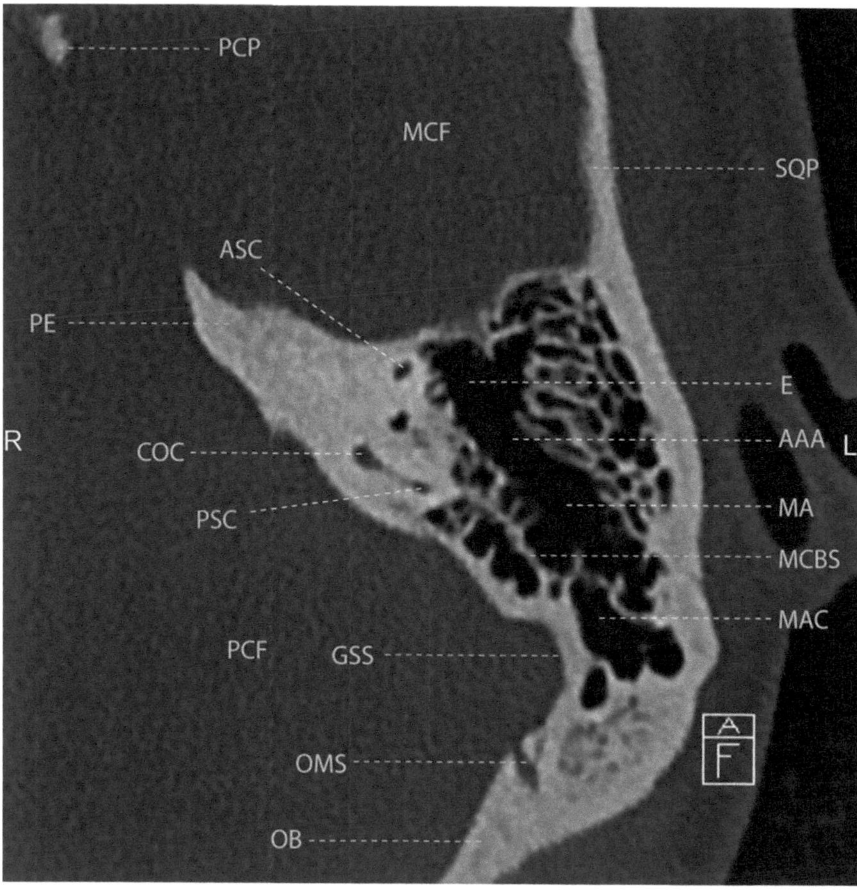

A10: HRCT axial reformat

AAA	aditus ad antrum	OB	occipital bone
ASC	anterior semicircular canal	OMS	occipitomastoid suture
COC	common osseous crus	PCF	posterior cranial fossa
E	epitympanum	PCP	posterior clinoid process
GSS	groove of the sigmoid sinus	PE	petrous part of the temporal bone
MA	mastoid antrum	PSC	posterior semicircular canal
MAC	mastoid air cells	SQP	squamous part of the temporal bone
MCBS	mastoid cells bony septa		
MCF	middle cranial fossa		

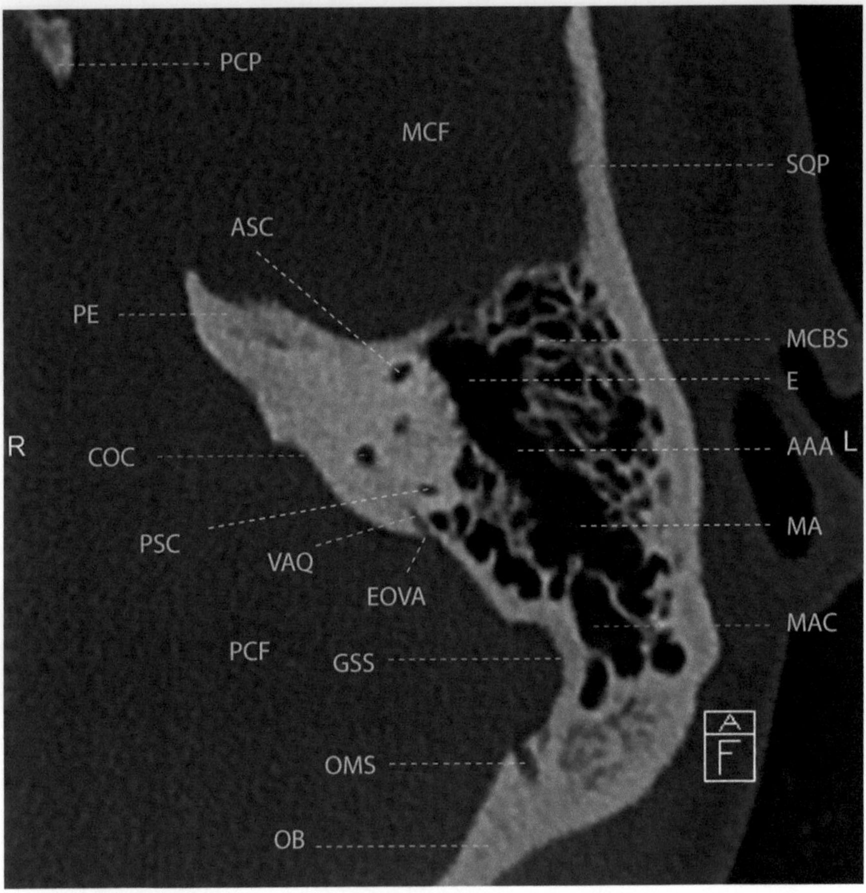

A11: HRCT axial reformat

AAA	aditus ad antrum
ASC	anterior semicircular canal
COC	common osseous crus
E	epitympanum
EOVA	external opening of the vestibular aqueduct
GSS	groove of the sigmoid sinus
MA	mastoid antrum
MAC	mastoid air cells
MCBS	mastoid cells bony septa
MCF	middle cranial fossa
OB	occipital bone
OMS	occipitomastoid suture
PCF	posterior cranial fossa
PCP	posterior clinoid process
PE	petrous part of the temporal bone
PSC	posterior semicircular canal
SQP	squamous part of the temporal bone
VAQ	vestibular aqueduct

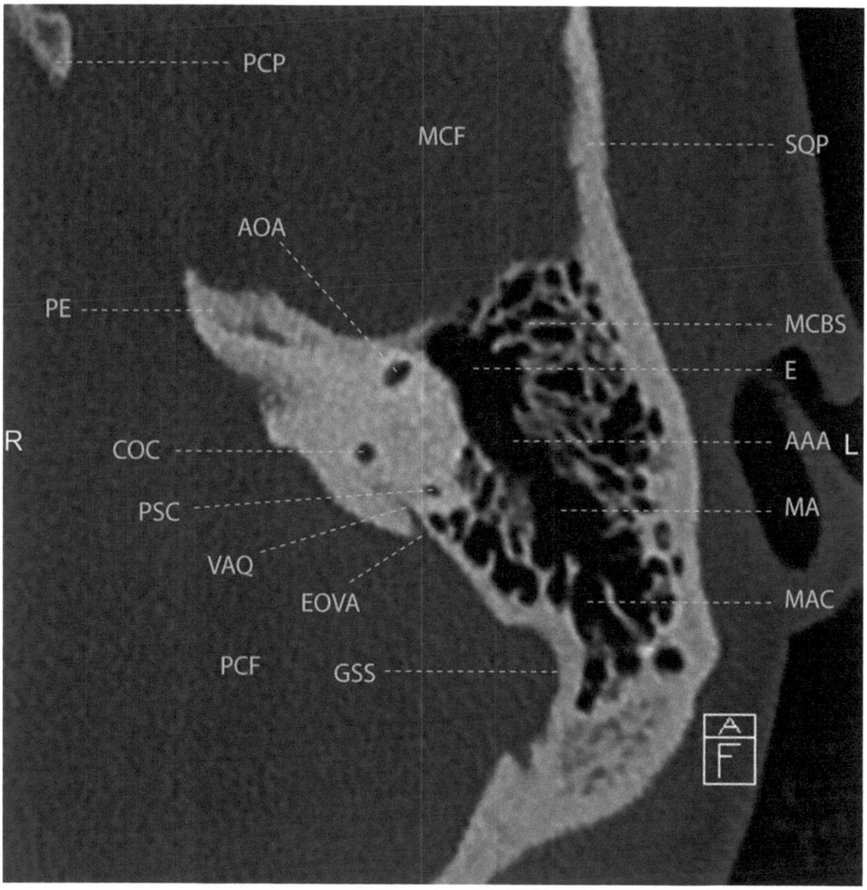

A12: HRCT axial reformat

AAA	aditus ad antrum
AOA	anterior osseous ampulla
COC	common osseous crus
E	epitympanum
EOVA	external opening of the vestibular aqueduct
GSS	groove of the sigmoid sinus
MA	mastoid antrum
MAC	mastoid air cells
MCBS	mastoid cells bony septa
MCF	middle cranial fossa
PCF	posterior cranial fossa
PCP	posterior clinoid process
PE	petrous part of the temporal bone
PSC	posterior semicircular canal
SQP	squamous part of the temporal bone
VAQ	vestibular aqueduct

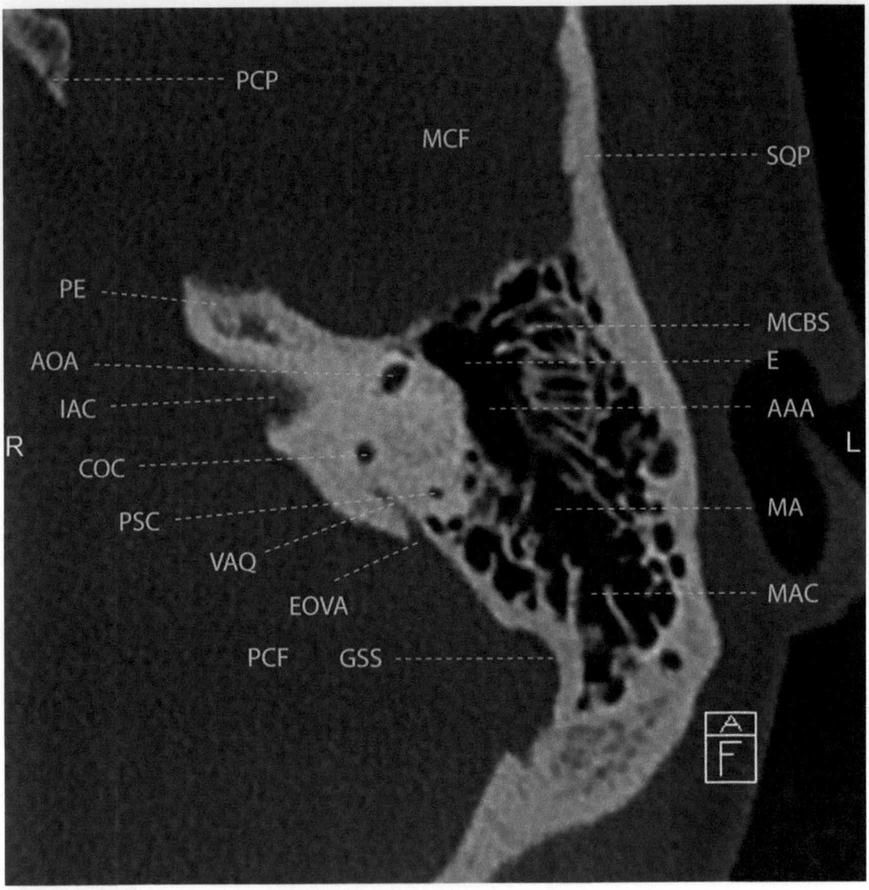

A13: HRCT axial reformat

AAA	aditus ad antrum	MCBS	mastoid cells bony septa
AOA	anterior osseous ampulla	MCF	middle cranial fossa
COC	common osseous crus	PCF	posterior cranial fossa
E	epitympanum	PCP	posterior clinoid process
EOVA	external opening of the vestibular aqueduct	PE	petrous part of the temporal bone
GSS	groove of the sigmoid sinus	PSC	posterior semicircular canal
IAC	internal auditory canal	SQP	squamous part of the temporal bone
MA	mastoid antrum		
MAC	mastoid air cells	VAQ	vestibular aqueduct

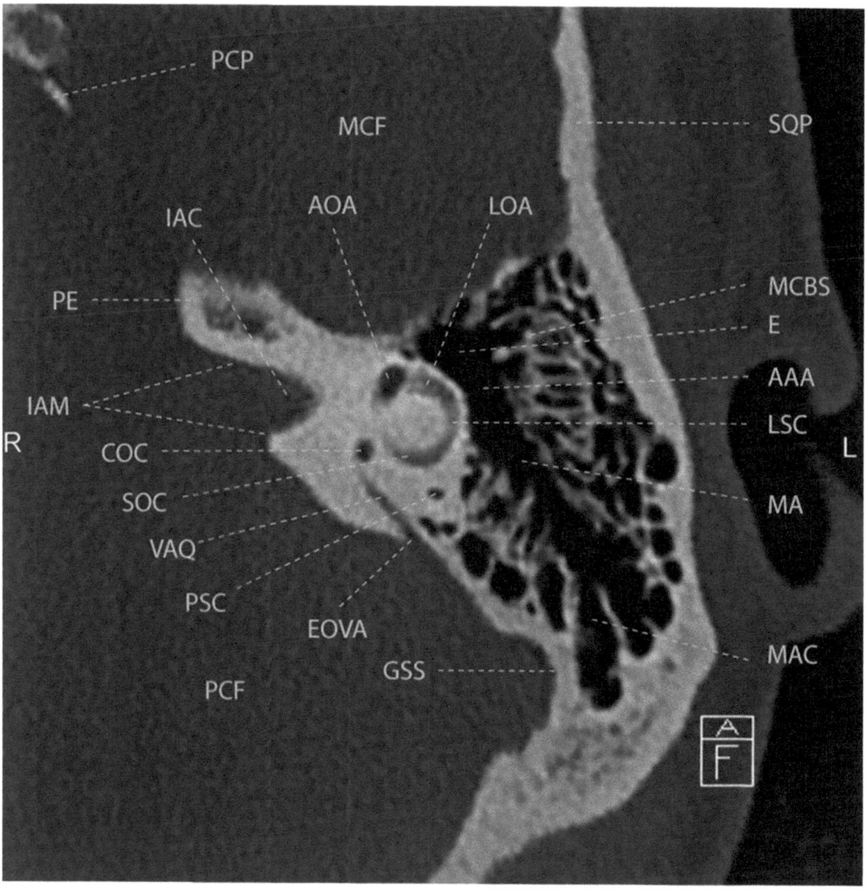

A14: HRCT axial reformat

AAA	aditus ad antrum	MAC	mastoid air cells
AOA	anterior osseous ampulla	MCBS	mastoid cells bony septa
COC	common osseous crus	MCF	middle cranial fossa
E	epitympanum	PCF	posterior cranial fossa
EOVA	external opening of the vestibular aqueduct	PCP	posterior clinoid process
GSS	groove of the sigmoid sinus	PE	petrous part of the temporal bone
IAC	internal auditory canal	PSC	posterior semicircular canal
IAM	internal auditory canal – meatus	SOC	simple osseous crus
LOA	lateral osseous ampulla	SQP	squamous part of the temporal bone
LSC	lateral semicircular canal	VAQ	vestibular aqueduct
MA	mastoid antrum		

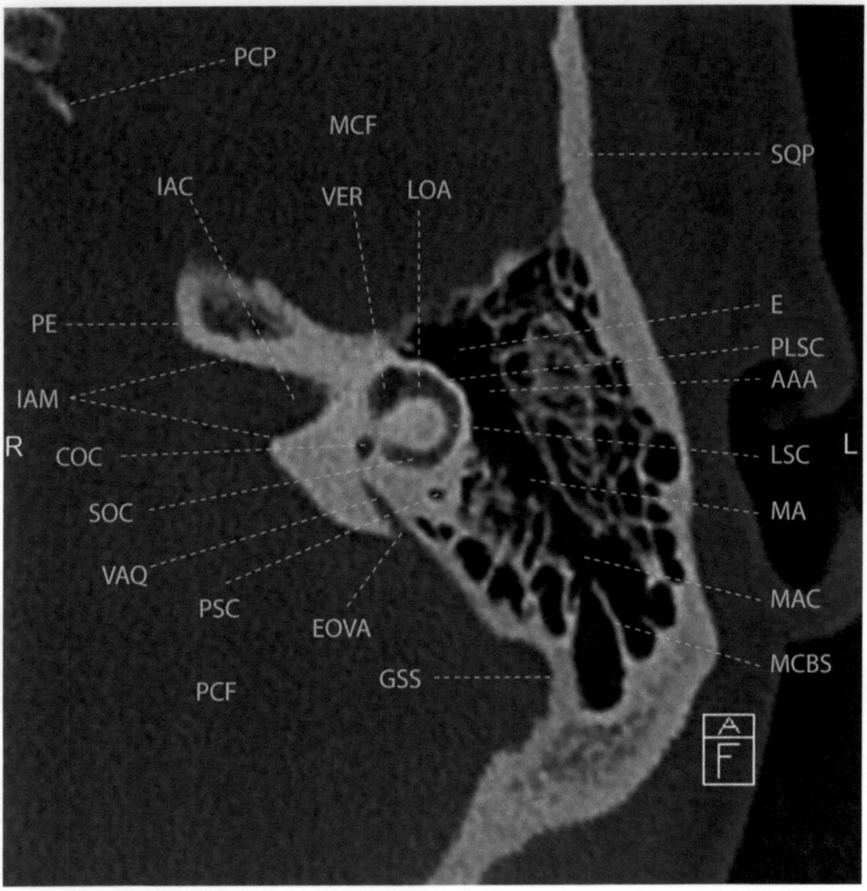

A15: HRCT axial reformat

AAA	aditus ad antrum	MCF	middle cranial fossa
COC	common osseous crus	PCF	posterior cranial fossa
E	epitympanum	PCP	posterior clinoid process
EOVA	external opening of the vestibular aqueduct	PLSC	prominence of the lateral semicircular canal
GSS	groove of the sigmoid sinus	PE	petrous part of the temporal bone
IAC	internal auditory canal		
IAM	internal auditory canal – meatus	PSC	posterior semicircular canal
LOA	lateral osseous ampulla	SOC	simple osseous crus
LSC	lateral semicircular canal	SQP	squamous part of the temporal bone
MA	mastoid antrum		
MAC	mastoid air cells	VAQ	vestibular aqueduct
MCBS	mastoid cells bony septa	VER	vestibule – elliptical recess

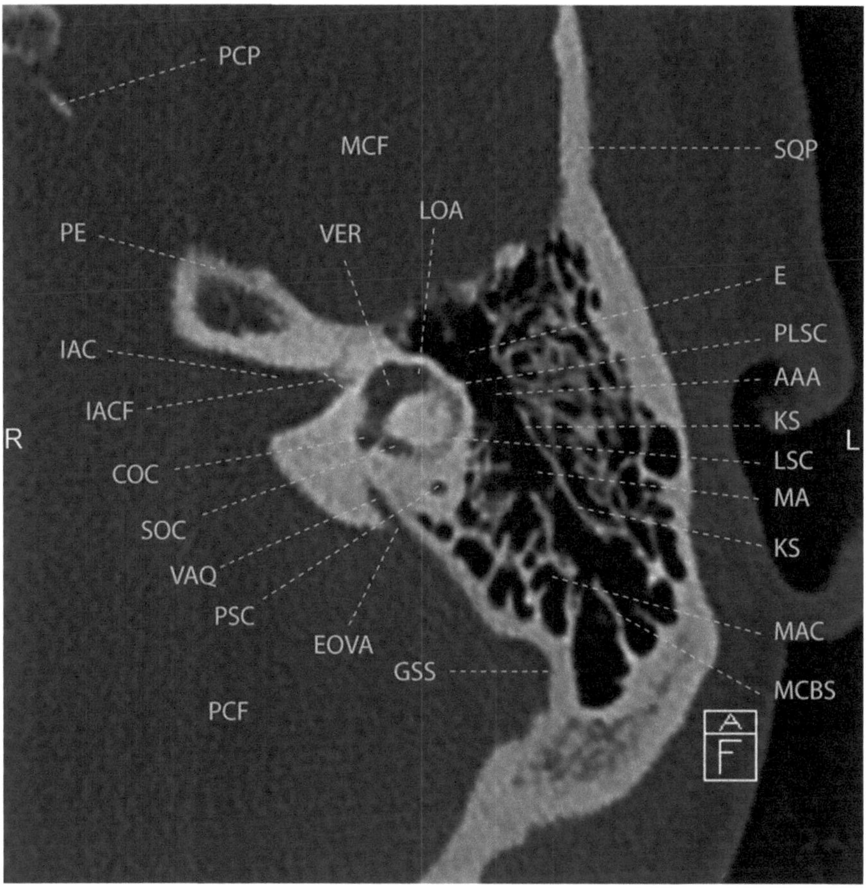

A16: HRCT axial reformat

AAA	aditus ad antrum
COC	common osseous crus
E	epitympanum
EOVA	external opening of the vestibular aqueduct
GSS	groove of the sigmoid sinus
IAC	internal auditory canal
IACF	internal auditory canal – fundus
KS	Koerner's septum
LOA	lateral osseous ampulla
LSC	lateral semicircular canal
MA	mastoid antrum
MAC	mastoid air cells
MCBS	mastoid cells bony septa
MCF	middle cranial fossa
PCF	posterior cranial fossa
PCP	posterior clinoid process
PLSC	prominence of the lateral semicircular canal
PE	petrous part of the temporal bone
PSC	posterior semicircular canal
SOC	simple osseous crus
SQP	squamous part of the temporal bone
VAQ	vestibular aqueduct
VER	vestibule – elliptical recess

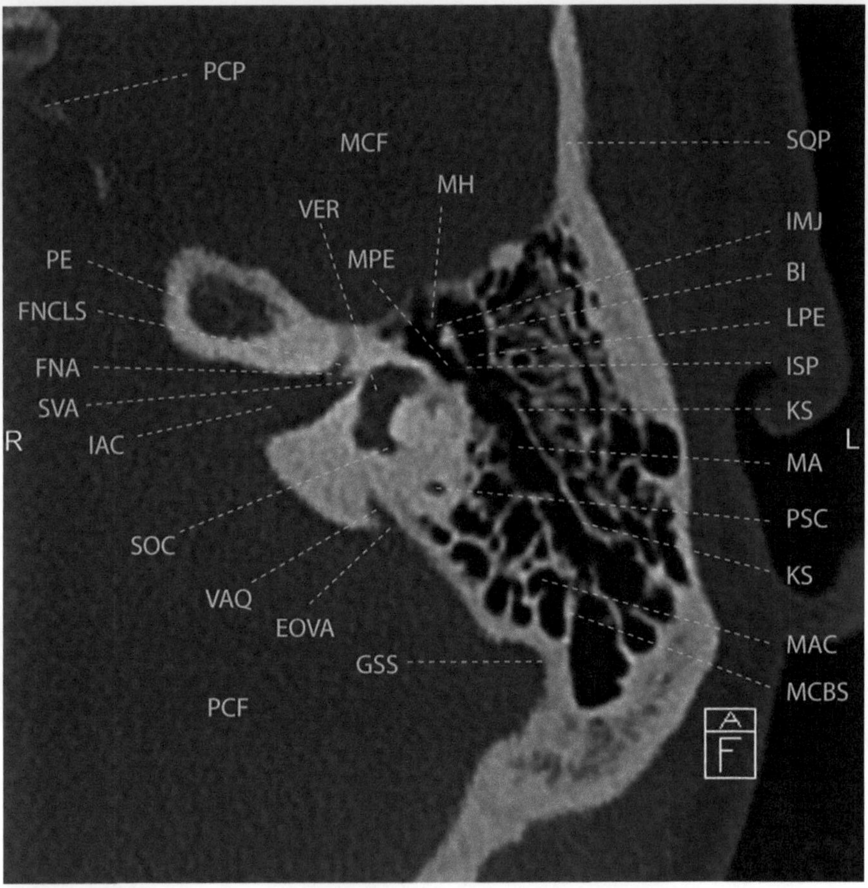

A17: HRCT axial reformat

BI	body of incus	MCBS	mastoid cells bony septa
EOVA	external opening of the vestibular aqueduct	MCF	middle cranial fossa
FNA	facial nerve area	MH	malleus – head
FNCLS	facial nerve canal – labyrinthine segment	MPE	medial posterior epitympanum
GSS	groove of the sigmoid sinus	PCF	posterior cranial fossa
IAC	internal auditory canal	PCP	posterior clinoid process
IMJ	incudomallear joint	PE	petrous part of the temporal bone
ISP	incus – short process	PSC	posterior semicircular canal
KS	Koerner's septum	SOC	simple osseous crus
LPE	lateral posterior epitympanum	SQP	squamous part of the temporal bone
MA	mastoid antrum	SVA	superior vestibular area
MAC	mastoid air cells	VAQ	vestibular aqueduct
		VER	vestibule – elliptical recess

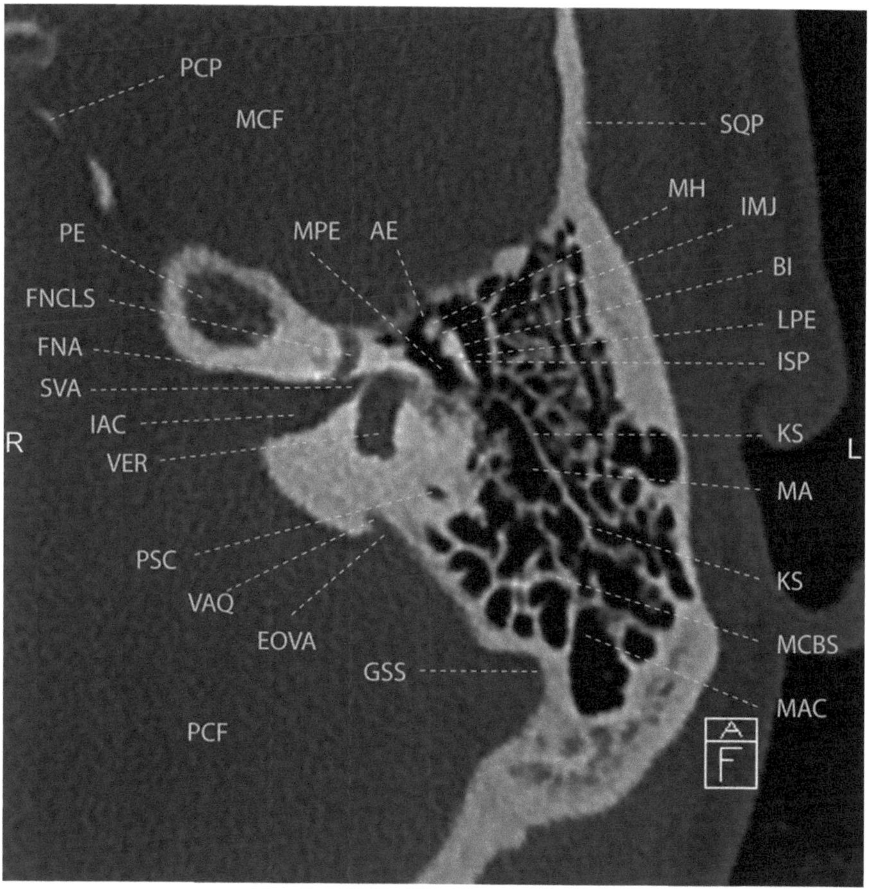

A18: HRCT axial reformat

AE	anterior epitympanum
BI	body of incus
EOVA	external opening of the vestibular aqueduct
FNA	facial nerve area
FNCLS	facial nerve canal – labyrinthine segment
GSS	groove of the sigmoid sinus
IAC	internal auditory canal
IMJ	incudomallear joint
ISP	incus – short process
KS	Koerner's septum
LPE	lateral posterior epitympanum
MA	mastoid antrum
MAC	mastoid air cells
MCBS	mastoid cells bony septa
MCF	middle cranial fossa
MH	malleus – head
MPE	medial posterior epitympanum
PCF	posterior cranial fossa
PCP	posterior clinoid process
PE	petrous part of the temporal bone
PSC	posterior semicircular canal
SQP	squamous part of the temporal bone
SVA	superior vestibular area
VAQ	vestibular aqueduct
VER	vestibule – elliptical recess

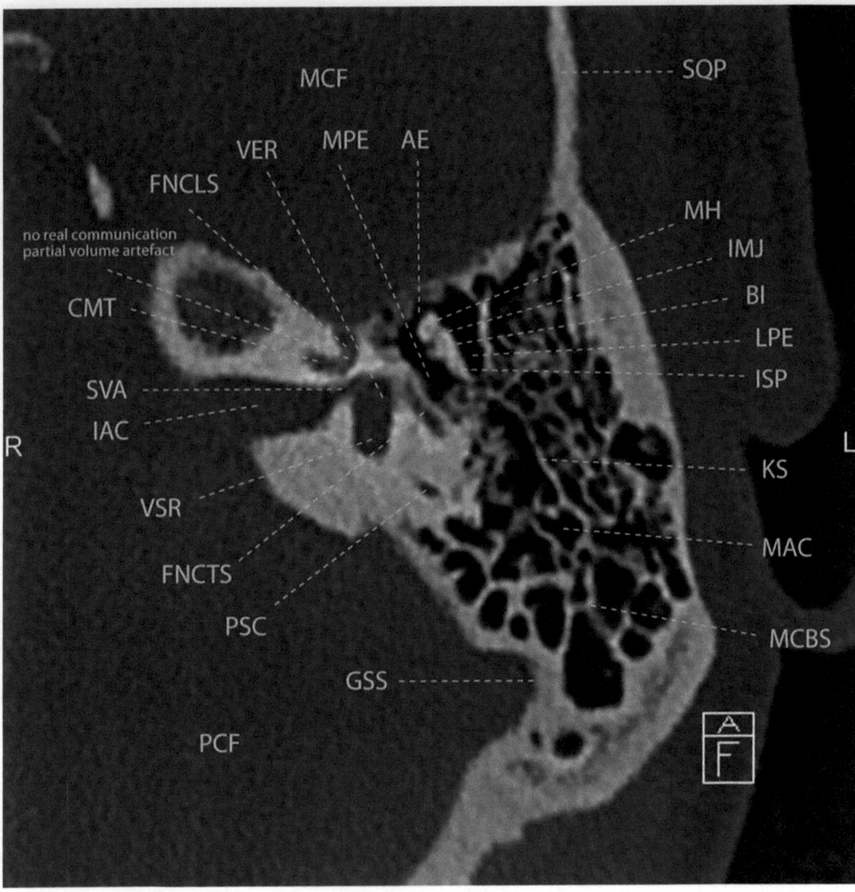

A19: HRCT axial reformat

AE	anterior epitympanum
BI	body of incus
CMT	cochlea – middle turn
FNCLS	facial nerve canal – labyrinthine segment
FNCTS	facial nerve canal – tympanic segment
GSS	groove of the sigmoid sinus
IAC	internal auditory canal
IMJ	incudomallear joint
ISP	incus – short process
KS	Koerner's septum
LPE	lateral posterior epitympanum
MAC	mastoid air cells
MCBS	mastoid cells bony septa
MCF	middle cranial fossa
MH	malleus – head
MPE	medial posterior epitympanum
PCF	posterior cranial fossa
PSC	posterior semicircular canal
SQP	squamous part of the temporal bone
SVA	superior vestibular area
VER	vestibule – elliptical recess
VSR	vestibule – spherical recess

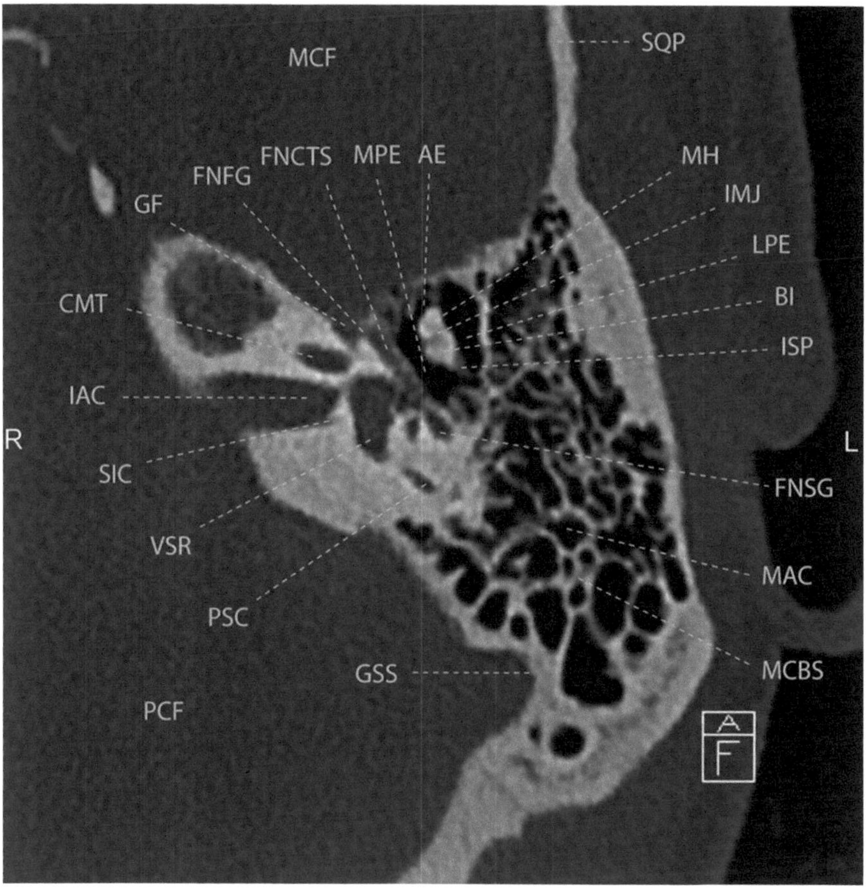

A20: HRCT axial reformat

AE	anterior epitympanum
BI	body of incus
CMT	cochlea – middle turn
FNCTS	facial nerve canal – tympanic segment
FNFG	facial nerve – first genu
FNSG	facial nerve – second genu
GF	geniculate fossa
GSS	groove of the sigmoid sinus
IAC	internal auditory canal
IMJ	incudomallear joint
ISP	incus – short process
LPE	lateral posterior epitympanum
MAC	mastoid air cells
MCBS	mastoid cells bony septa
MCF	middle cranial fossa
MH	malleus – head
MPE	medial posterior epitympanum
PCF	posterior cranial fossa
PSC	posterior semicircular canal
SIC	singular canal
SQP	squamous part of the temporal bone
VSR	vestibule – spherical recess

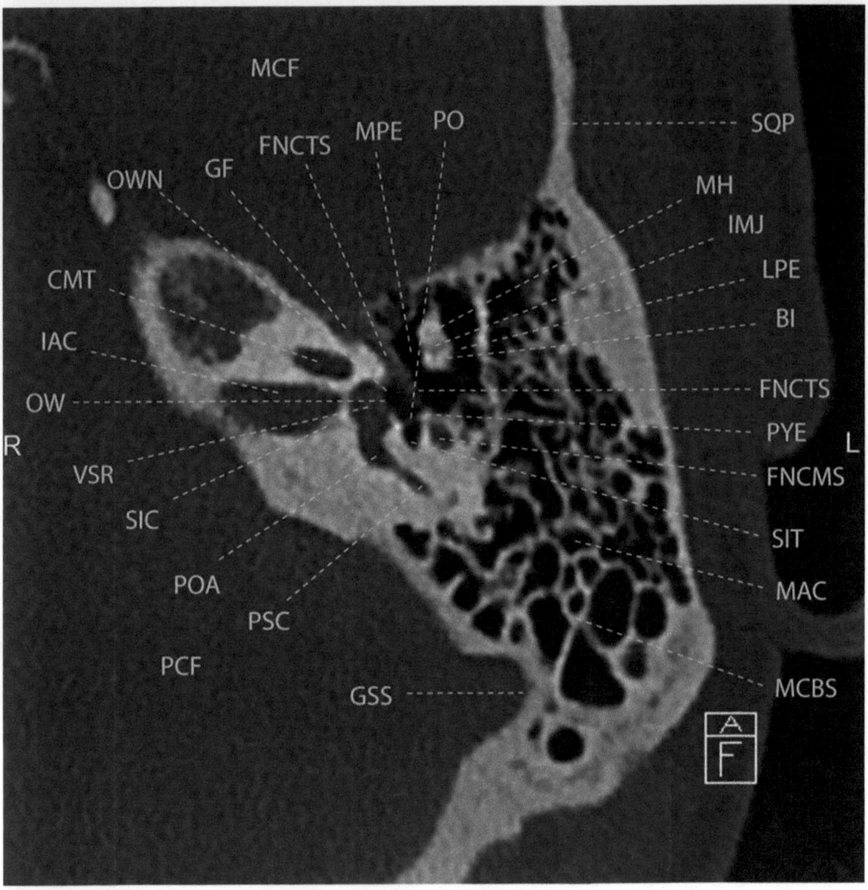

A21: HRCT axial reformat

BI	body of incus
CMT	cochlea – middle turn
FNCMS	facial nerve canal – mastoid segment
FNCTS	facial nerve canal – tympanic segment
GF	geniculate fossa
GSS	groove of the sigmoid sinus
IAC	internal auditory canal
IMJ	incudomallear joint
LPE	lateral posterior epitympanum
MAC	mastoid air cells
MCBS	mastoid cells bony septa
MCF	middle cranial fossa
MH	malleus – head
MPE	medial posterior epitympanum
OW	oval window
OWN	oval window niche
PCF	posterior cranial fossa
PO	ponticulus
POA	posterior osseous ampulla
PSC	posterior semicircular canal
PYE	pyramidal eminence
SIC	singular canal
SIT	sinus tympani
SQP	squamous part of the temporal bone
VSR	vestibule – spherical recess

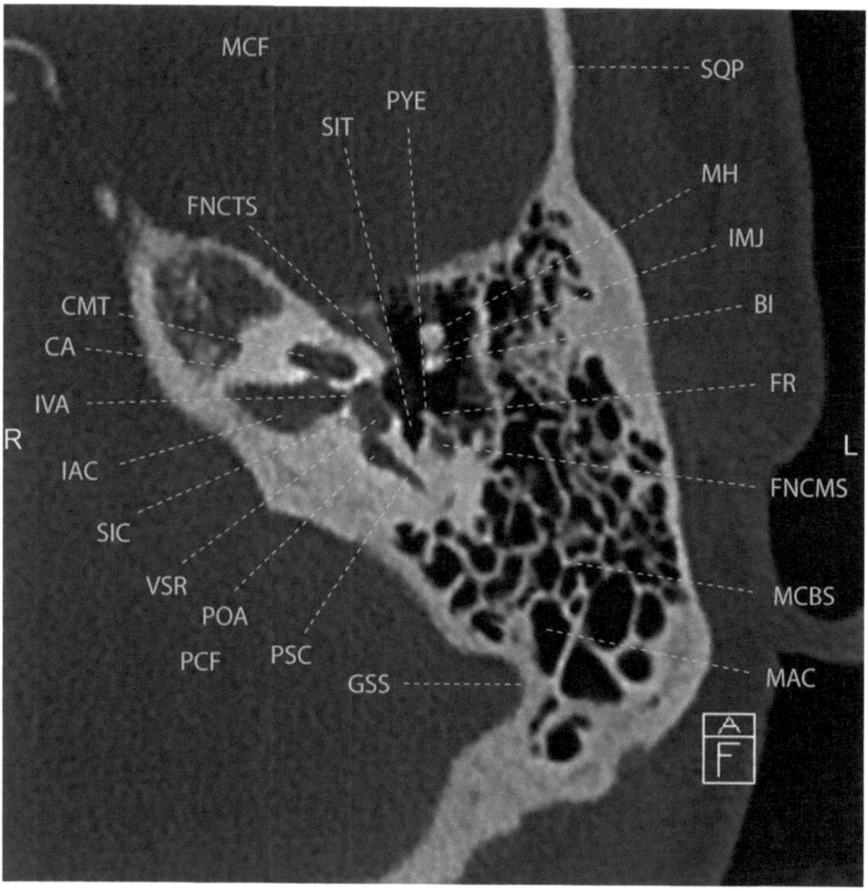

A22: HRCT axial reformat

BI	body of incus	MCBS	mastoid cells bony septa
CA	cochlear area	MCF	middle cranial fossa
CMT	cochlea – middle turn	MH	malleus – head
FNCMS	facial nerve canal – mastoid segment	PCF	posterior cranial fossa
		POA	posterior osseous ampulla
FNCTS	facial nerve canal – tympanic segment	PSC	posterior semicircular canal
		PYE	pyramidal eminence
FR	facial recess	SIC	singular canal
GSS	groove of the sigmoid sinus	SIT	sinus tympani
IAC	internal auditory canal	SQP	squamous part of the temporal bone
IMJ	incudomallear joint		
IVA	inferior vestibular area	VSR	vestibule – spherical recess
MAC	mastoid air cells		

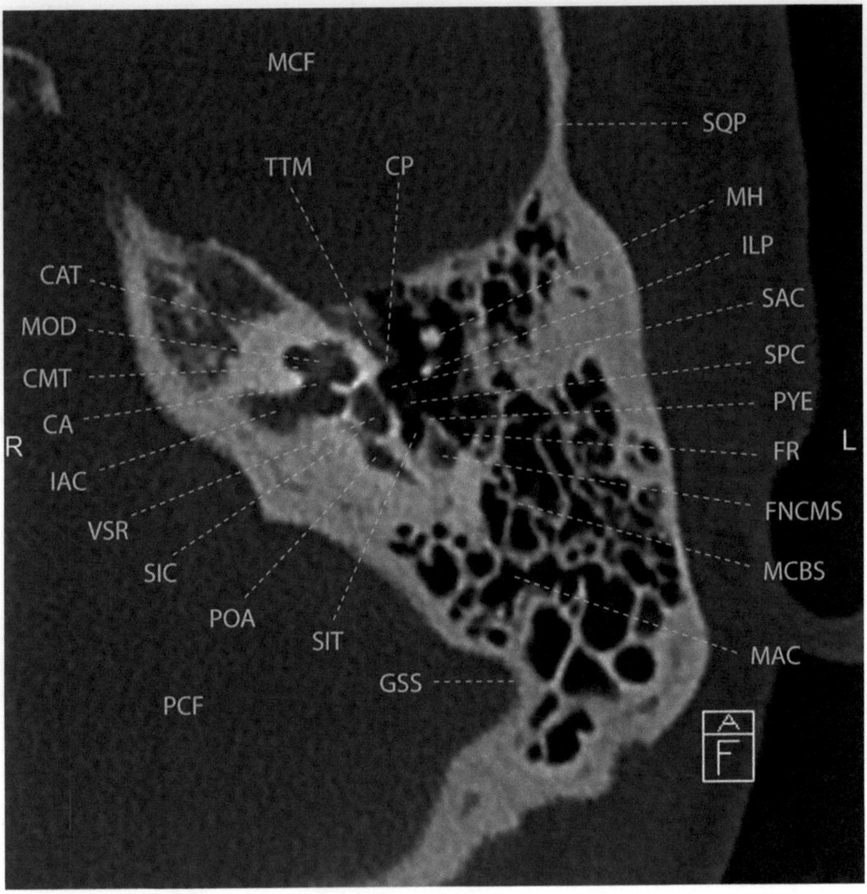

A23: HRCT axial reformat

CA	cochlear area
CAT	cochlea – apical turn
CMT	cochlea – middle turn
CP	cochleariform process
FNCMS	facial nerve canal – mastoid segment
FR	facial recess
GSS	groove of the sigmoid sinus
IAC	internal auditory canal
ILP	incus – long process
MAC	mastoid air cells
MCBS	mastoid cells bony septa
MCF	middle cranial fossa
MH	malleus – head
MOD	modiolus
PCF	posterior cranial fossa
POA	posterior osseous ampulla
PYE	pyramidal eminence
SAC	stapes – anterior crus
SIC	singular canal
SIT	sinus tympani
SPC	stapes – posterior crus
SQP	squamous part of the temporal bone
TTM	tensor tympani muscle
VSR	vestibule – spherical recess

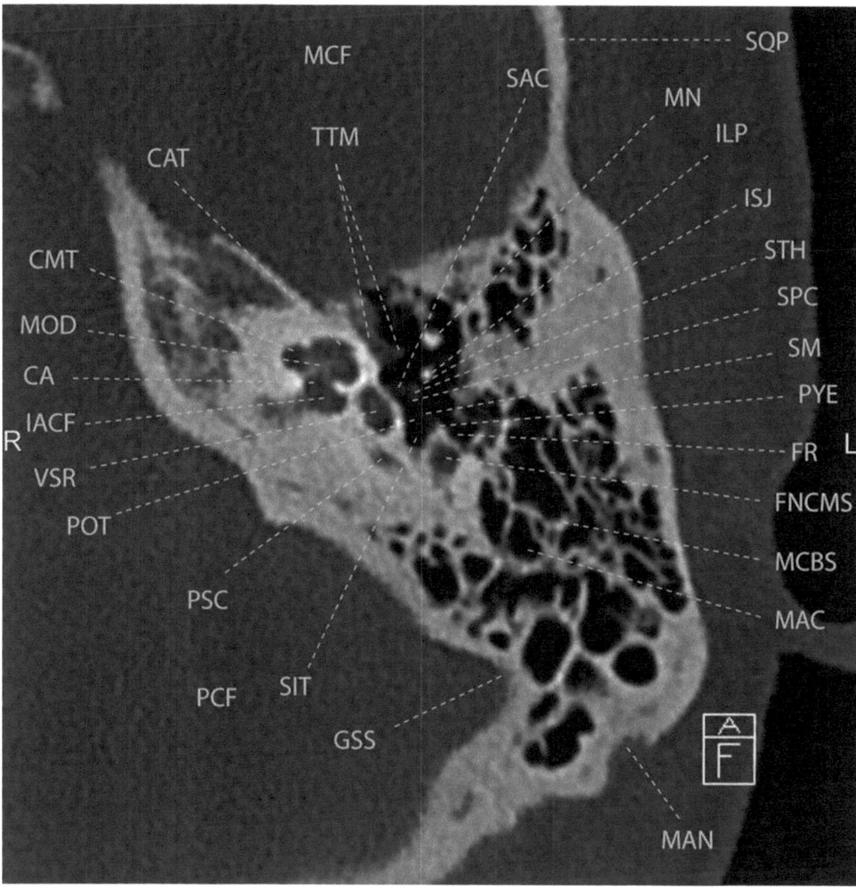

A24: HRCT axial reformat

CA	cochlear area	MN	malleus – neck
CAT	cochlea – apical turn	MOD	modiolus
CMT	cochlea – middle turn	PCF	posterior cranial fossa
FNCMS	facial nerve canal – mastoid segment	POT	posterior tympanum
		PSC	posterior semicircular canal
FR	facial recess	PYE	pyramidal eminence
GSS	groove of the sigmoid sinus	SAC	stapes – anterior crus
IACF	internal auditory canal – fundus	SIT	sinus tympani
		SM	stapedius muscle
ILP	incus – long process	SPC	stapes – posterior crus
ISJ	incudostapedial joint	SQP	squamous part of the temporal bone
MAC	mastoid air cells		
MAN	mastoid notch	STH	stapes – head
MCBS	mastoid cells bony septa	TTM	tensor tympani muscle
MCF	middle cranial fossa	VSR	vestibule – spherical recess

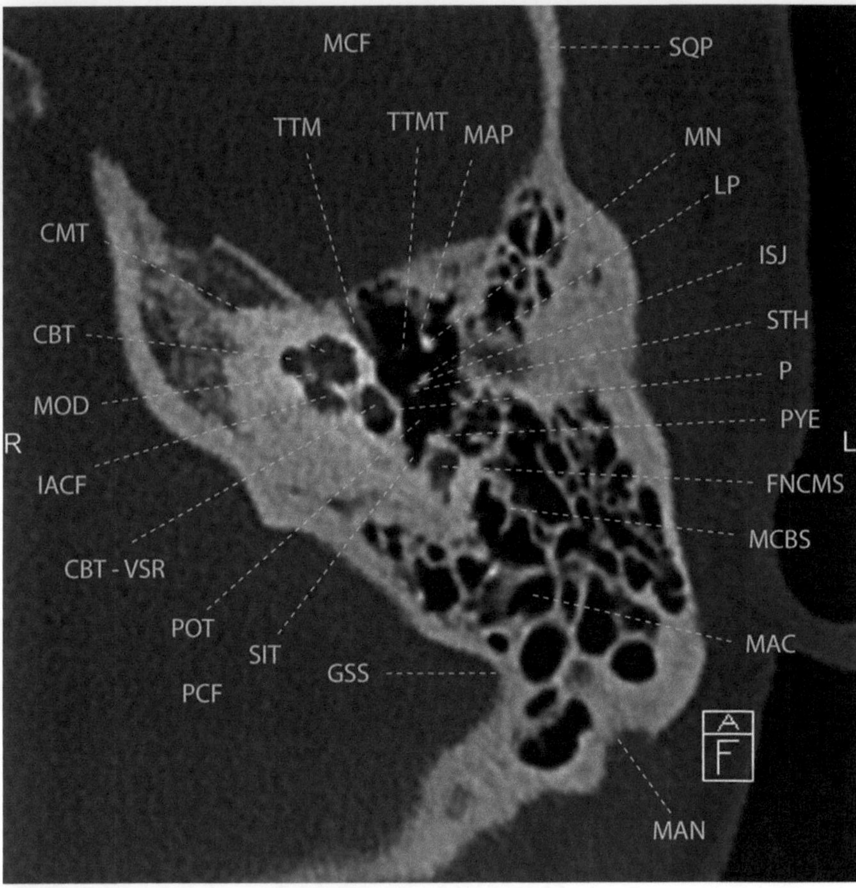

A25: HRCT axial reformat

CBT	cochlea – basilar turn	MN	malleus – neck
CMT	cochlea – middle turn	MOD	modiolus
FNCMS	facial nerve canal – mastoid segment	P	promontory
		PCF	posterior cranial fossa
GSS	groove of the sigmoid sinus	POT	posterior tympanum
IACF	internal auditory canal – fundus	PYE	pyramidal eminence
		SIT	sinus tympani
ISJ	incudostapedial joint	SQP	squamous part of the temporal bone
LP	lenticular process		
MAC	mastoid air cells	STH	stapes – head
MAN	mastoid notch	TTM	tensor tympani muscle
MAP	malleus – anterior process	TTMT	tensor tympani muscle – tendon
MCBS	mastoid cells bony septa		
MCF	middle cranial fossa	VSR	vestibule – spherical recess

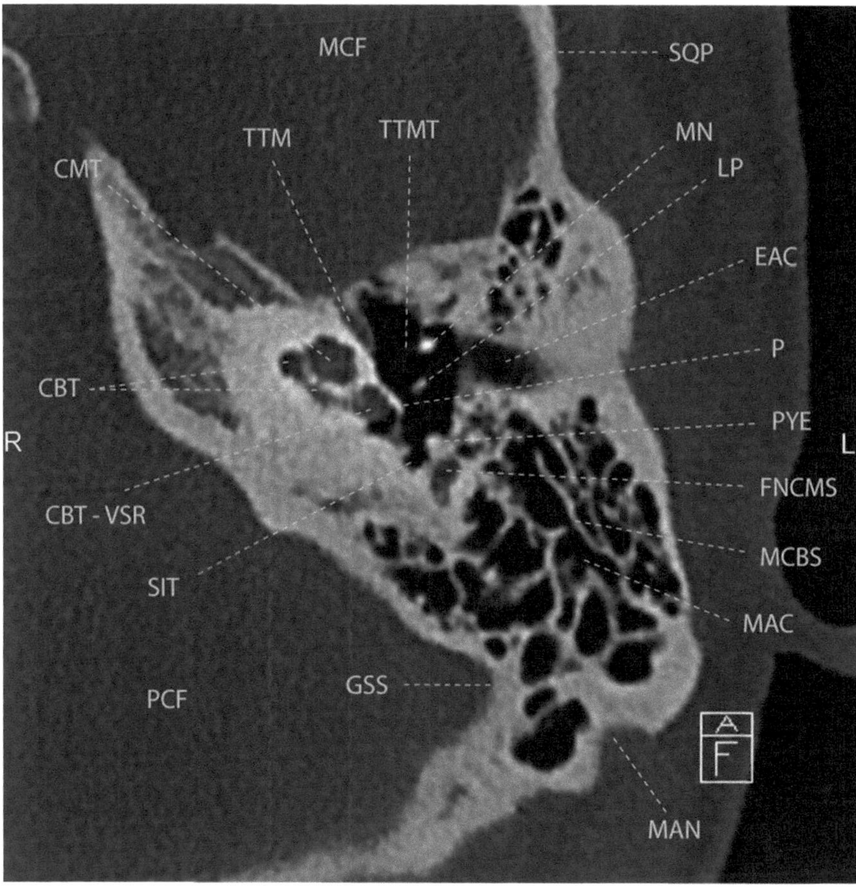

A26: HRCT axial reformat

CBT	cochlea – basilar turn	MN	malleus – neck
CMT	cochlea – middle turn	P	promontory
EAC	external auditory canal	PCF	posterior cranial fossa
FNCMS	facial nerve canal – mastoid segment	PYE	pyramidal eminence
		SIT	sinus tympani
GSS	groove of the sigmoid sinus	SQP	squamous part of the temporal bone
LP	lenticular process		
MAC	mastoid air cells	TTM	tensor tympani muscle
MAN	mastoid notch	TTMT	tensor tympani muscle – tendon
MCBS	mastoid cells bony septa		
MCF	middle cranial fossa	VSR	vestibule – spherical recess

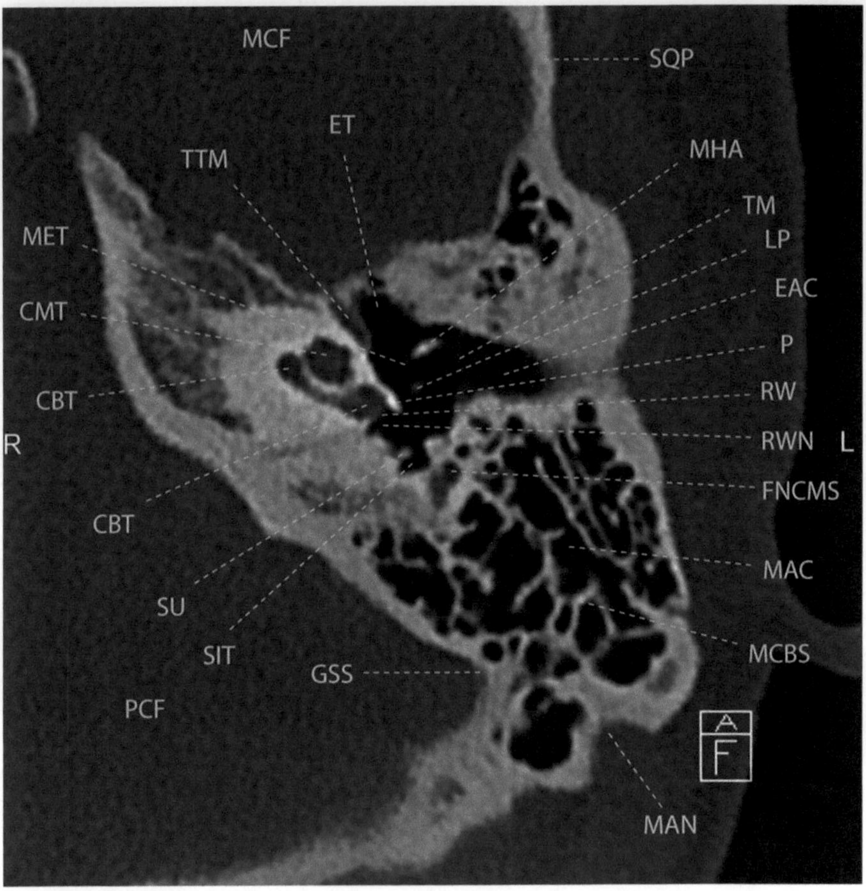

A27: HRCT axial reformat

CBT	cochlea – basilar turn	MET	mesotympanum
CMT	cochlea – middle turn	MHA	malleus – handle
EAC	external auditory canal	P	promontory
ET	Eustachian tube	PCF	posterior cranial fossa
FNCMS	facial nerve canal – mastoid segment	RW	round window
		RWN	round window niche
GSS	groove of the sigmoid sinus	SIT	sinus tympani
LP	lenticular process	SQP	squamous part of the temporal bone
MAC	mastoid air cells		
MAN	mastoid notch	SU	subiculum
MCBS	mastoid cells bony septa	TM	tympanic membrane
MCF	middle cranial fossa	TTM	tensor tympani muscle

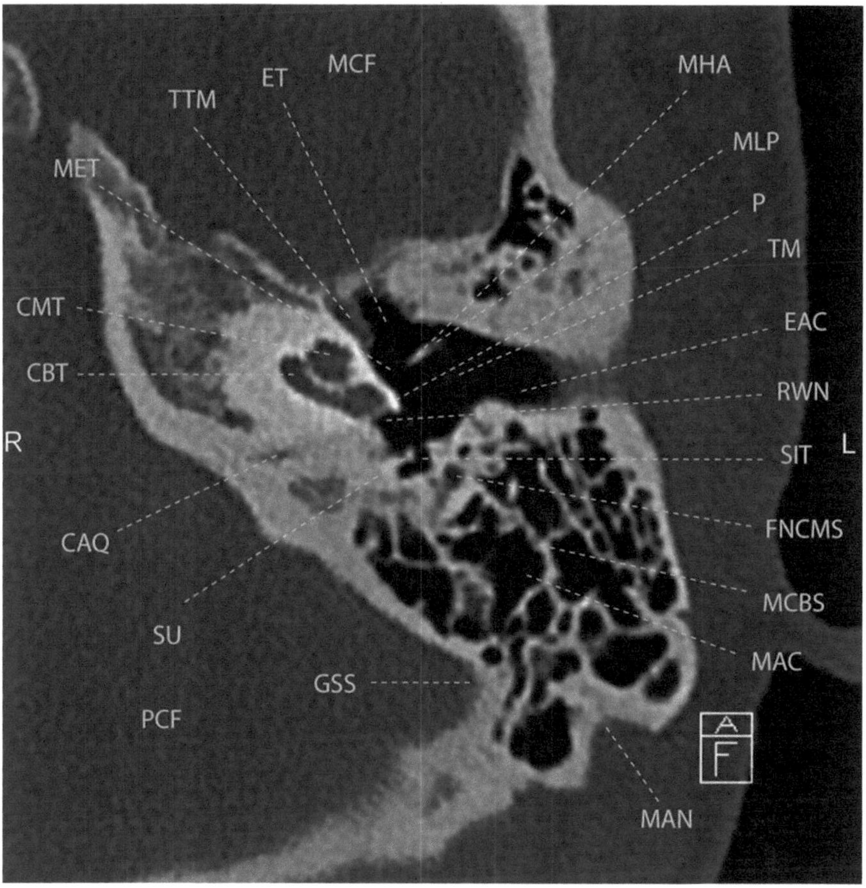

A28: HRCT axial reformat

CAQ	cochlear aqueduct	MCF	middle cranial fossa
CBT	cochlea – basilar turn	MET	mesotympanum
CMT	cochlea – middle turn	MHA	malleus – handle
EAC	external auditory canal	MLP	malleus – lateral process
ET	Eustachian tube	P	promontory
FNCMS	facial nerve canal – mastoid segment	PCF	posterior cranial fossa
		RWN	round window niche
GSS	groove of the sigmoid sinus	SIT	sinus tympani
MAC	mastoid air cells	SU	subiculum
MAN	mastoid notch	TM	tympanic membrane
MCBS	mastoid cells bony septa	TTM	tensor tympani muscle

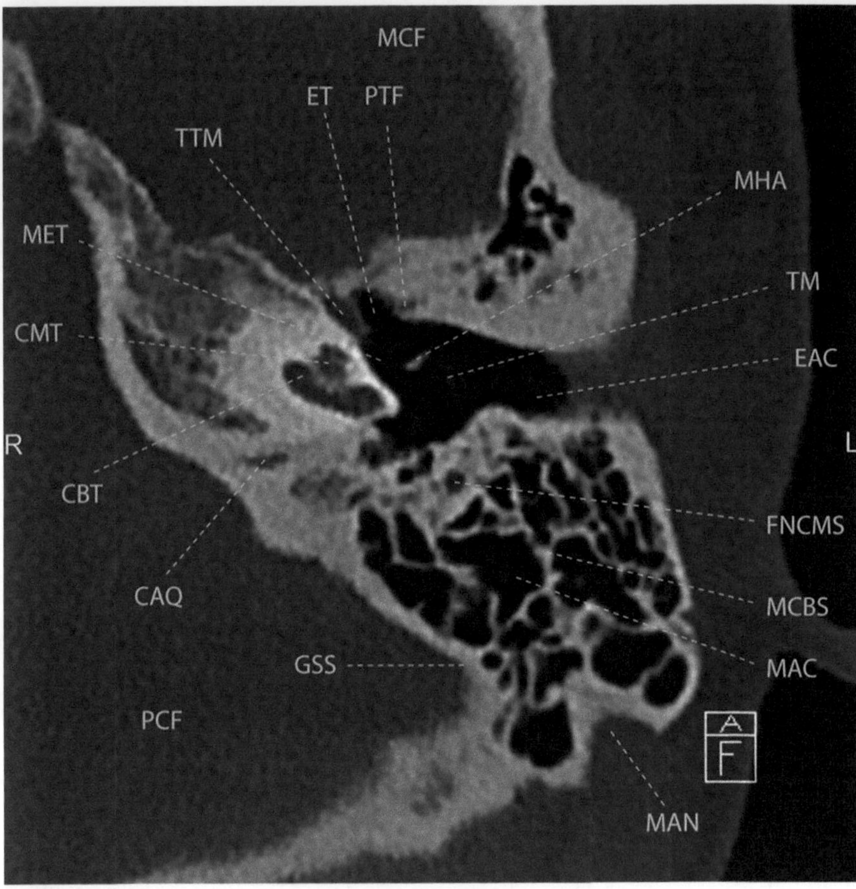

A29: HRCT axial reformat

CAQ	cochlear aqueduct	MAN	mastoid notch
CBT	cochlea – basilar turn	MCBS	mastoid cells bony septa
CMT	cochlea – middle turn	MCF	middle cranial fossa
EAC	external auditory canal	MET	mesotympanum
ET	Eustachian tube	MHA	malleus – handle
FNCMS	facial nerve canal – mastoid segment	PCF	posterior cranial fossa
		PTF	petrotympanic fissure
GSS	groove of the sigmoid sinus	TM	tympanic membrane
MAC	mastoid air cells	TTM	tensor tympani muscle

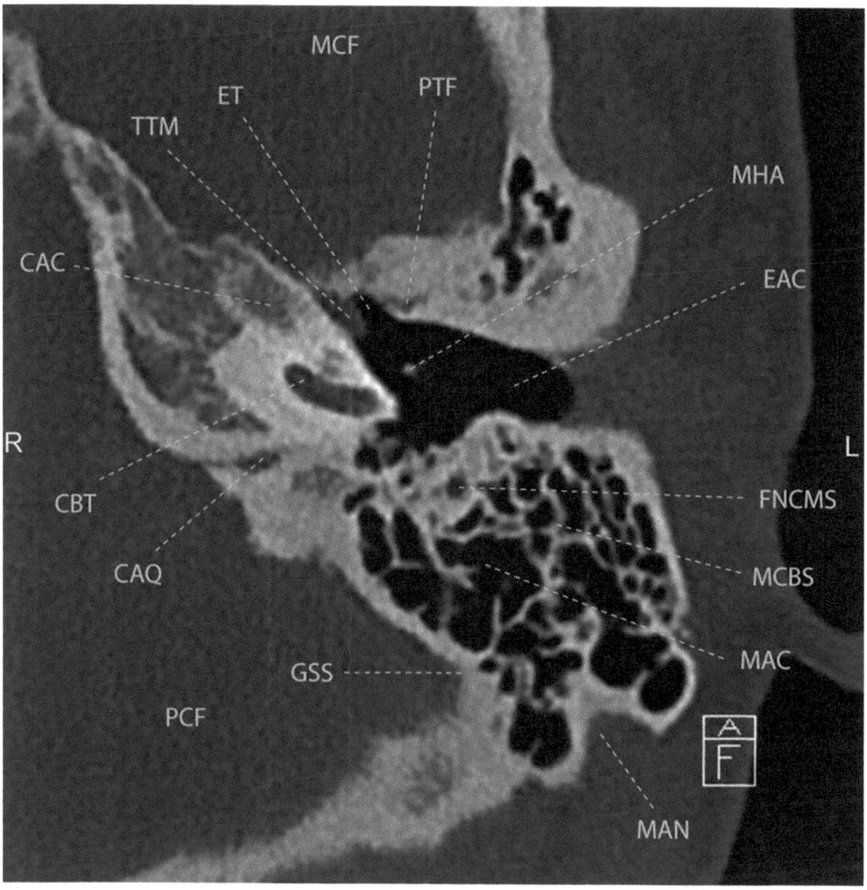

A30: HRCT axial reformat

CAC	carotid canal	MAC	mastoid air cells
CAQ	cochlear aqueduct	MAN	mastoid notch
CBT	cochlea – basilar turn	MCBS	mastoid cells bony septa
EAC	external auditory canal	MCF	middle cranial fossa
ET	Eustachian tube	MHA	malleus – handle
FNCMS	facial nerve canal – mastoid segment	PCF	posterior cranial fossa
		PTF	petrotympanic fissure
GSS	groove of the sigmoid sinus	TTM	tensor tympani muscle

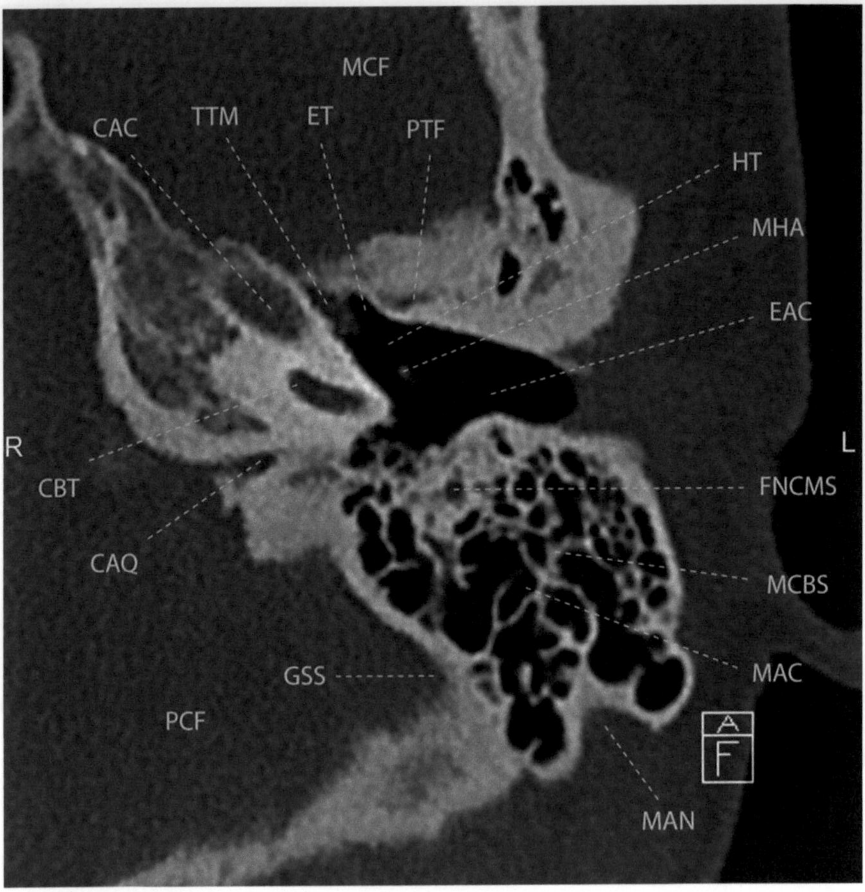

A31: HRCT axial reformat

CAC	carotid canal
CAQ	cochlear aqueduct
CBT	cochlea – basilar turn
EAC	external auditory canal
ET	Eustachian tube
FNCMS	facial nerve canal – mastoid segment
GSS	groove of the sigmoid sinus
HT	hypotympanum
MAC	mastoid air cells
MAN	mastoid notch
MCBS	mastoid cells bony septa
MCF	middle cranial fossa
MHA	malleus – handle
PCF	posterior cranial fossa
PTF	petrotympanic fissure
TTM	tensor tympani muscle

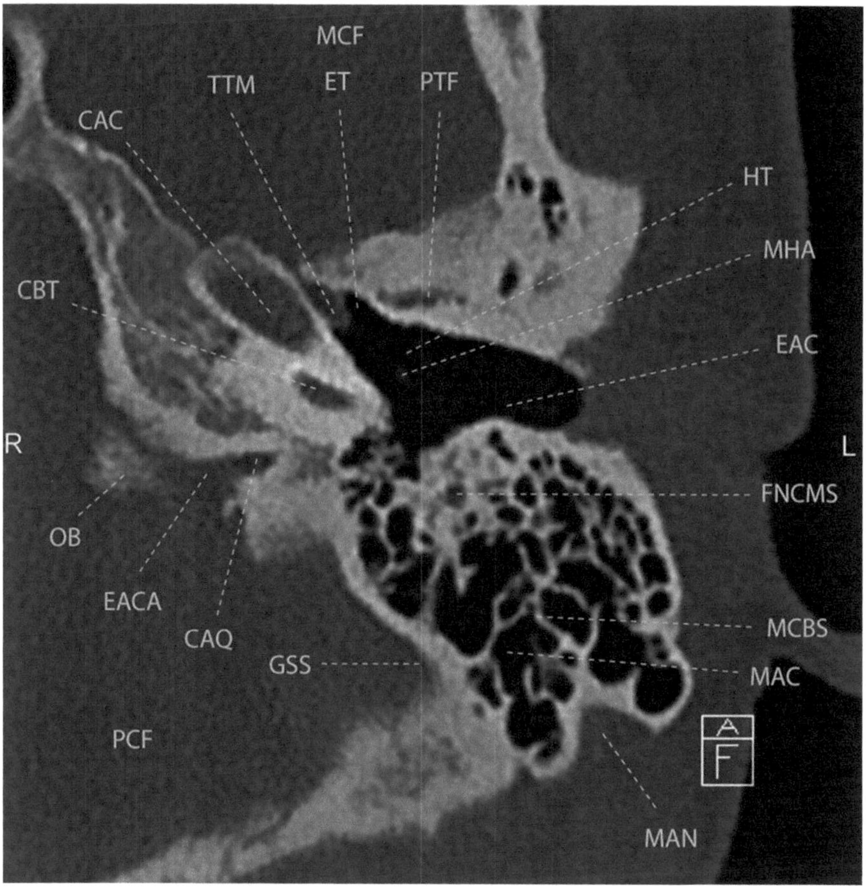

A32: HRCT axial reformat

CAC	carotid canal	HT	hypotympanum
CAQ	cochlear aqueduct	MAC	mastoid air cells
CBT	cochlea – basilar turn	MAN	mastoid notch
EAC	external auditory canal	MCBS	mastoid cells bony septa
EACA	external aperture of the cochlear aqueduct	MCF	middle cranial fossa
		MHA	malleus – handle
ET	Eustachian tube	OB	occipital bone
FNCMS	facial nerve canal – mastoid segment	PCF	posterior cranial fossa
		PTF	petrotympanic fissure
GSS	groove of the sigmoid sinus	TTM	tensor tympani muscle

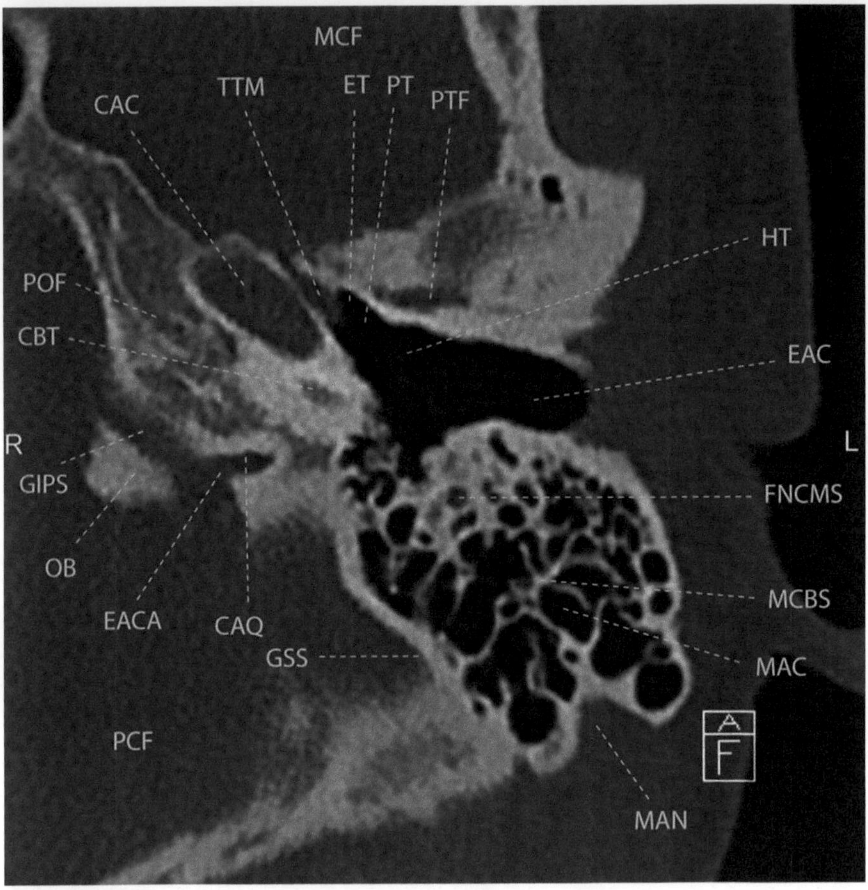

A33: HRCT axial reformat

CAC	carotid canal	HT	hypotympanum
CAQ	cochlear aqueduct	MAC	mastoid air cells
CBT	cochlea – basilar turn	MAN	mastoid notch
EAC	external auditory canal	MCBS	mastoid cells bony septa
EACA	external aperture of the cochlear aqueduct	MCF	middle cranial fossa
		OB	occipital bone
ET	Eustachian tube	PCF	posterior cranial fossa
FNCMS	facial nerve canal – mastoid segment	POF	petrooccipital fissure
		PT	protympanum
GIPS	groove of the inferior petrosal sinus	PTF	petrotympanic fissure
		TTM	tensor tympani muscle
GSS	groove of the sigmoid sinus		

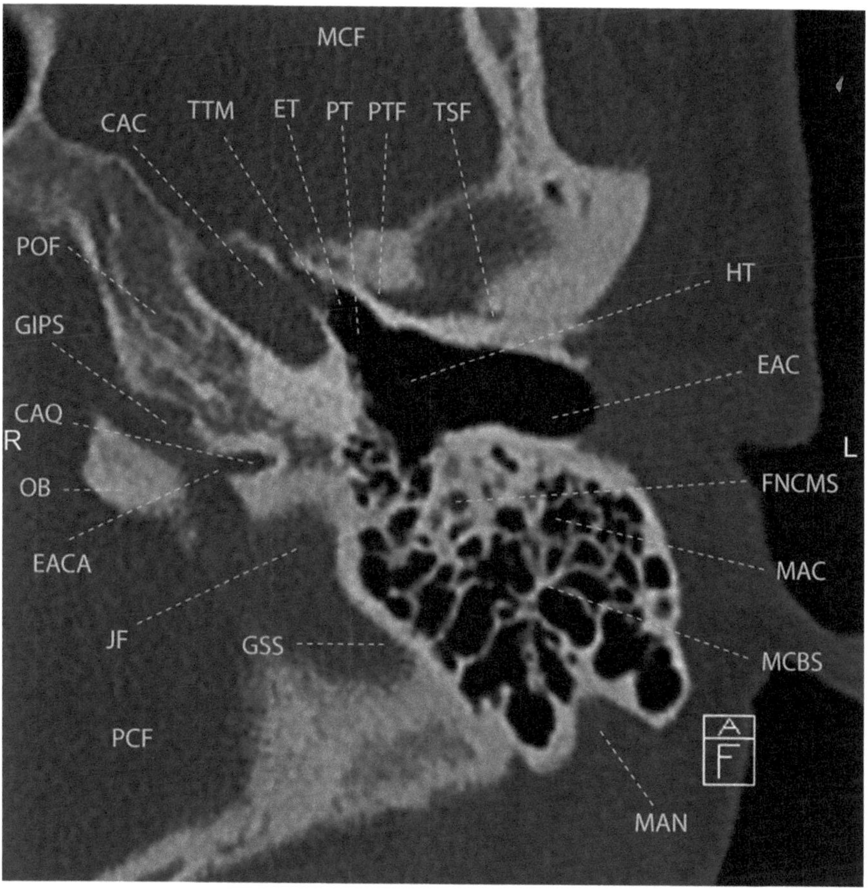

A34: HRCT axial reformat

CAC	carotid canal	JF	jugular foramen
CAQ	cochlear aqueduct	MAC	mastoid air cells
EAC	external auditory canal	MAN	mastoid notch
EACA	external aperture of the cochlear aqueduct	MCBS	mastoid cells bony septa
		MCF	middle cranial fossa
ET	Eustachian tube	OB	occipital bone
FNCMS	facial nerve canal – mastoid segment	PCF	posterior cranial fossa
		POF	petrooccipital fissure
GIPS	groove of the inferior petrosal sinus	PT	protympanum
		PTF	petrotympanic fissure
GSS	groove of the sigmoid sinus	TSF	tympanosquamous fissure
HT	hypotympanum	TTM	tensor tympani muscle

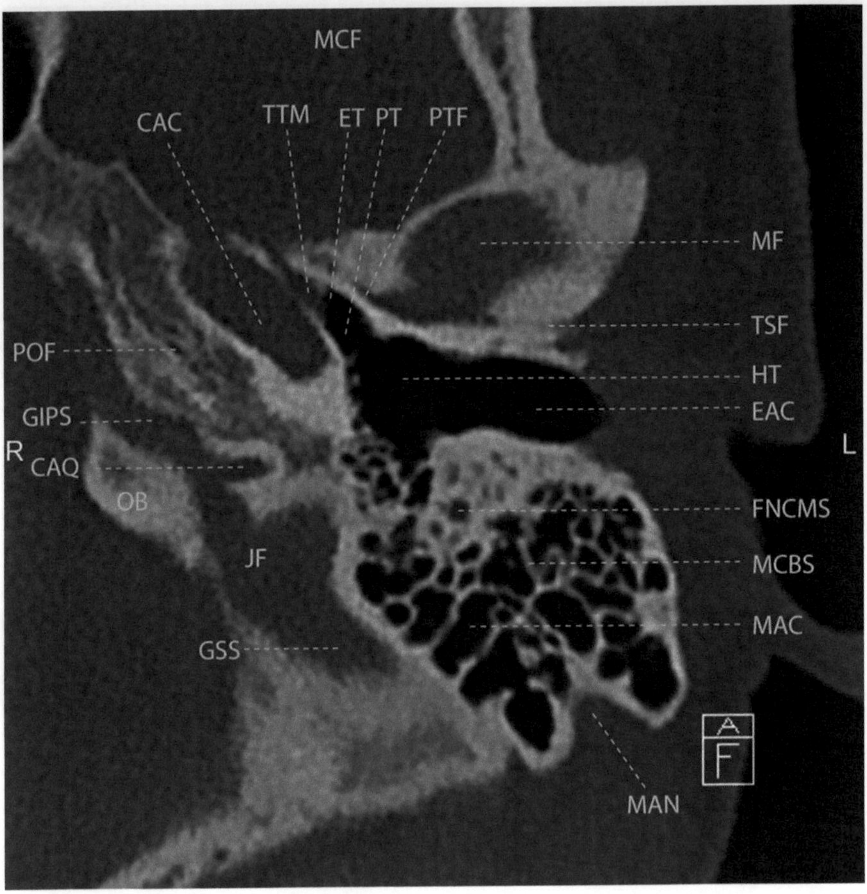

A35: HRCT axial reformat

CAC	carotid canal
CAQ	cochlear aqueduct
EAC	external auditory canal
ET	Eustachian tube
FNCMS	facial nerve canal – mastoid segment
GIPS	groove of the inferior petrosal sinus
GSS	groove of the sigmoid sinus
HT	hypotympanum
JF	jugular foramen
MAC	mastoid air cells
MAN	mastoid notch
MCBS	mastoid cells bony septa
MCF	middle cranial fossa
MF	mandibular fossa
OB	occipital bone
POF	petrooccipital fissure
PT	protympanum
PTF	petrotympanic fissure
TSF	tympanosquamous fissure
TTM	tensor tympani muscle

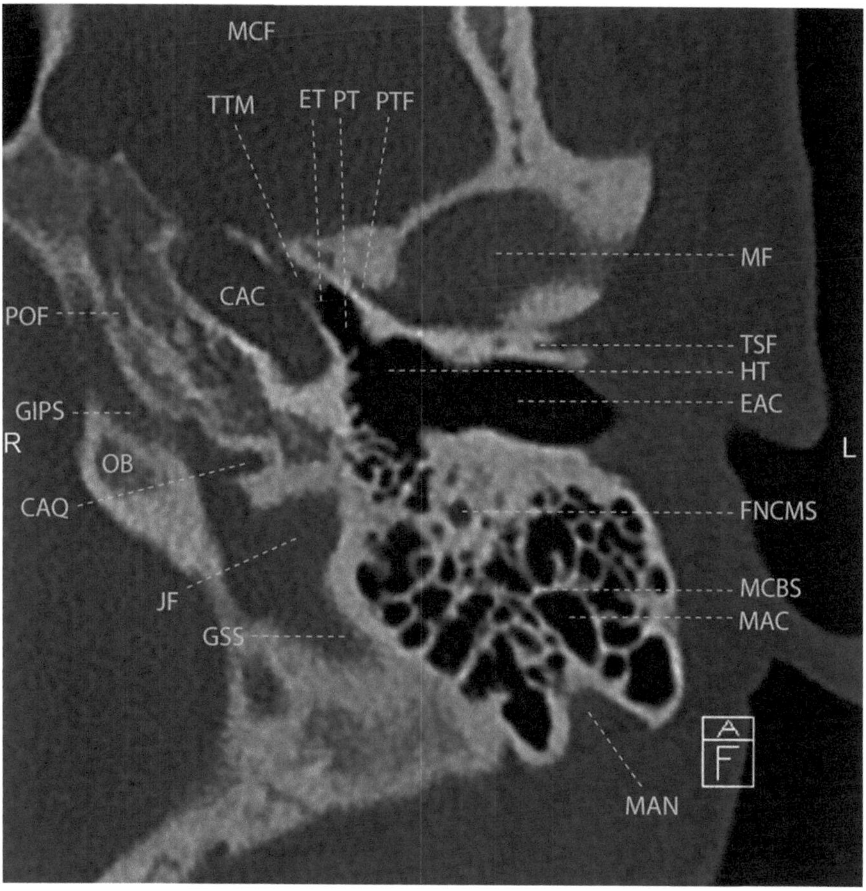

A36: HRCT axial reformat

CAC	carotid canal	MAC	mastoid air cells
CAQ	cochlear aqueduct	MAN	mastoid notch
EAC	external auditory canal	MCBS	mastoid cells bony septa
ET	Eustachian tube	MCF	middle cranial fossa
FNCMS	facial nerve canal – mastoid segment	MF	mandibular fossa
		OB	occipital bone
GIPS	groove of the inferior petrosal sinus	POF	petrooccipital fissure
		PT	protympanum
GSS	groove of the sigmoid sinus	PTF	petrotympanic fissure
HT	hypotympanum	TSF	tympanosquamous fissure
JF	jugular foramen	TTM	tensor tympani muscle

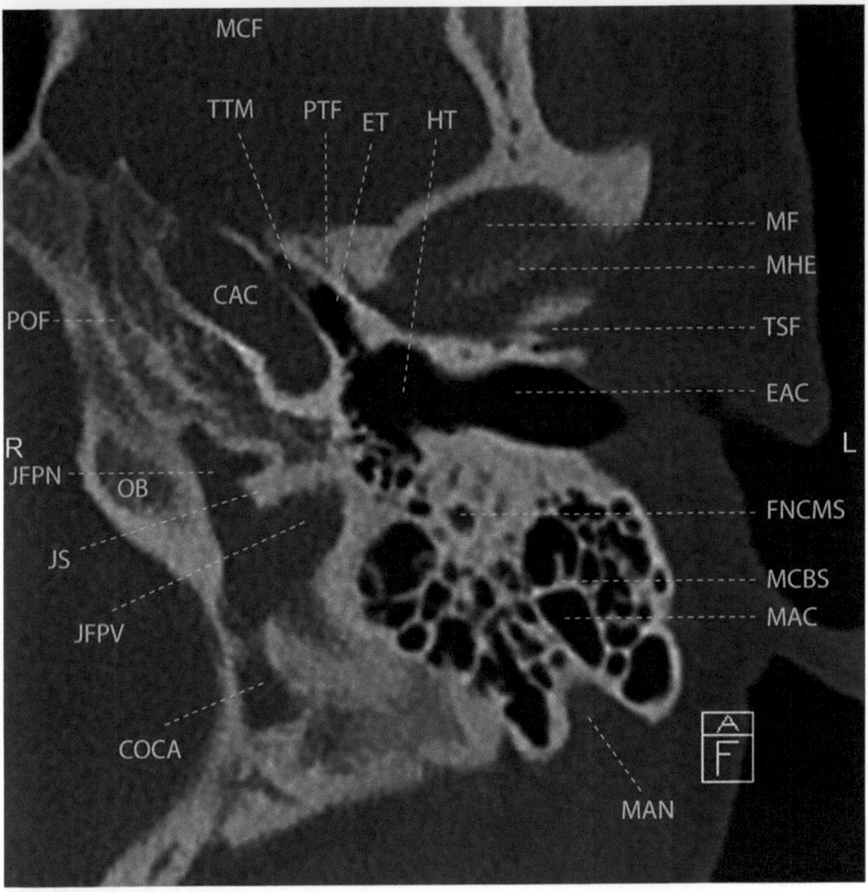

A37: HRCT axial reformat

CAC	carotid canal	MAC	mastoid air cells
COCA	condylar canal	MAN	mastoid notch
EAC	external auditory canal	MCBS	mastoid cells bony septa
ET	Eustachian tube	MCF	middle cranial fossa
FNCMS	facial nerve canal – mastoid segment	MF	mandibular fossa
		MHE	mandible – head
HT	hypotympanum	OB	occipital bone
JFPN	jugular foramen – pars nervosa	POF	petrooccipital fissure
		PTF	petrotympanic fissure
JFPV	jugular foramen – pars vascularis	TSF	tympanosquamous fissure
		TTM	tensor tympani muscle
JS	jugular spine		

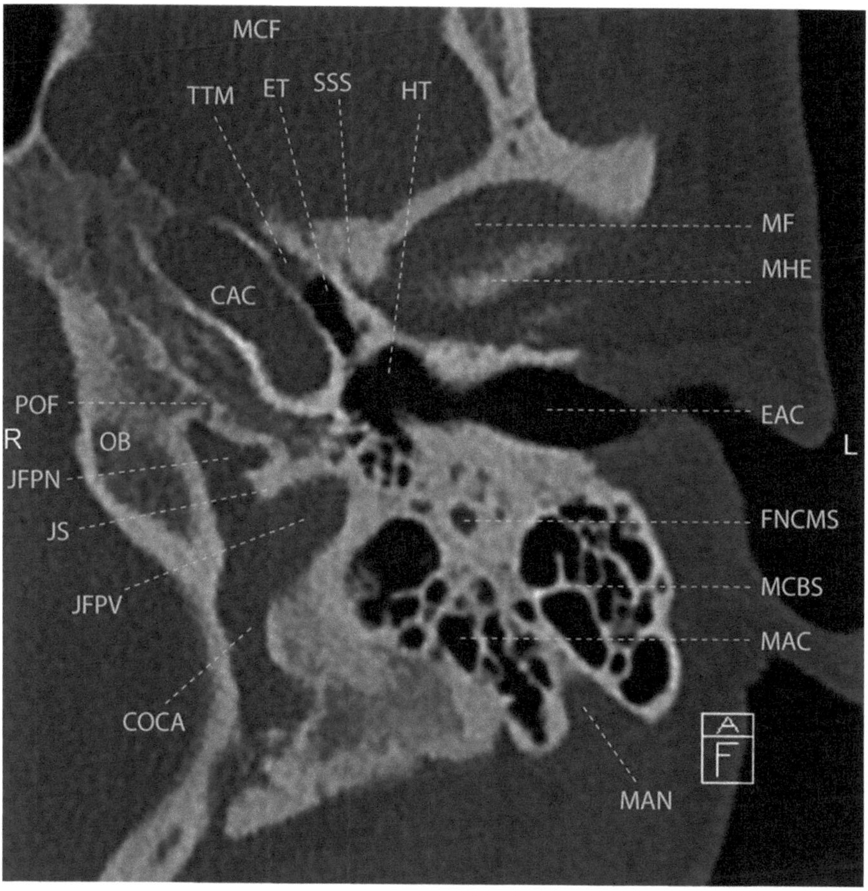

A38: HRCT axial reformat

CAC	carotid canal	JS	jugular spine
COCA	condylar canal	MAC	mastoid air cells
EAC	external auditory canal	MAN	mastoid notch
ET	Eustachian tube	MCBS	mastoid cells bony septa
FNCMS	facial nerve canal – mastoid segment	MCF	middle cranial fossa
HT	hypotympanum	MF	mandibular fossa
JFPN	jugular foramen – pars nervosa	MHE	mandible – head
JFPV	jugular foramen – pars vascularis	OB	occipital bone
		POF	petrooccipital fissure
		SSS	sphenosquamosal suture
		TTM	tensor tympani muscle

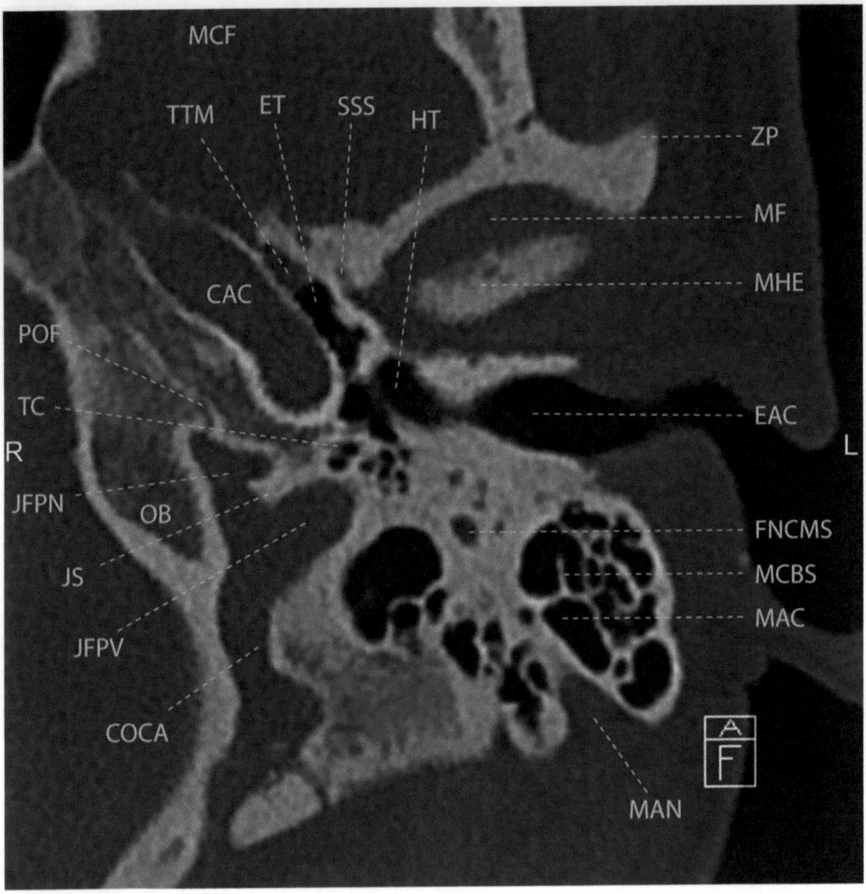

A39: HRCT axial reformat

CAC	carotid canal	MAC	mastoid air cells
COCA	condylar canal	MAN	mastoid notch
EAC	external auditory canal	MCBS	mastoid cells bony septa
ET	Eustachian tube	MCF	middle cranial fossa
FNCMS	facial nerve canal – mastoid segment	MF	mandibular fossa
HT	hypotympanum	MHE	mandible – head
JFPN	jugular foramen – pars nervosa	OB	occipital bone
JFPV	jugular foramen – pars vascularis	POF	petrooccipital fissure
JS	jugular spine	SSS	sphenosquamosal suture
		TC	tympanic canaliculus
		TTM	tensor tympani muscle
		ZP	zygomatic process

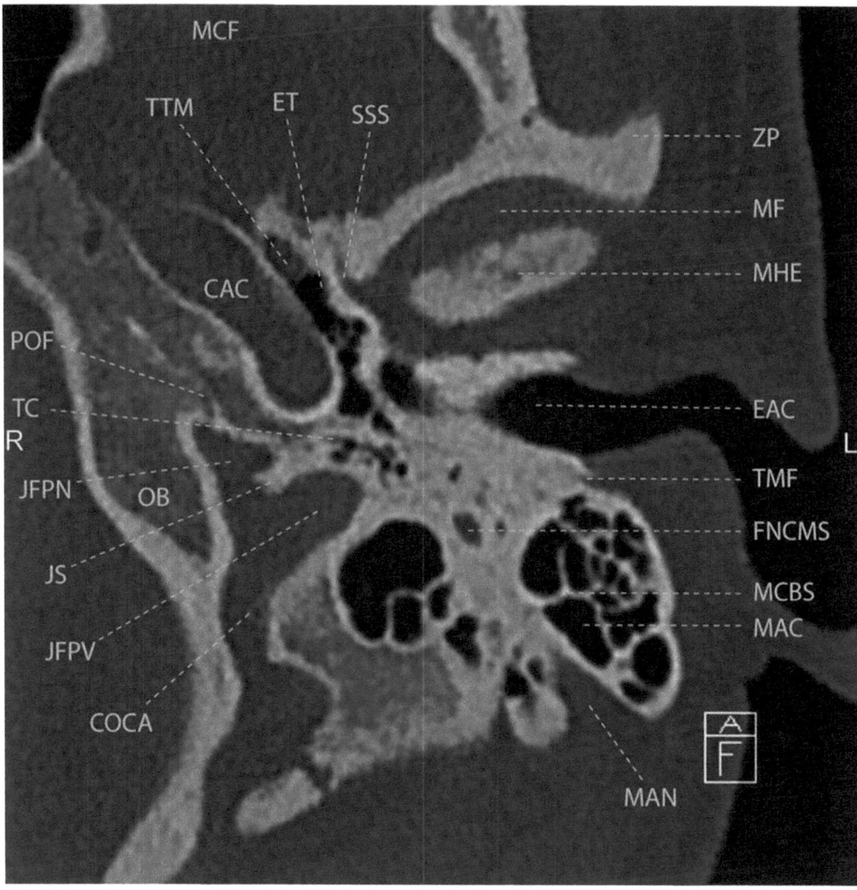

A40: HRCT axial reformat

CAC	carotid canal	MAN	mastoid notch
COCA	condylar canal	MCBS	mastoid cells bony septa
EAC	external auditory canal	MCF	middle cranial fossa
ET	Eustachian tube	MF	mandibular fossa
FNCMS	facial nerve canal – mastoid segment	MHE	mandible – head
		OB	occipital bone
JFPN	jugular foramen – pars nervosa	POF	petrooccipital fissure
		SSS	sphenosquamosal suture
JFPV	jugular foramen – pars vascularis	TC	tympanic canaliculus
		TMF	tympanomastoid fissure
JS	jugular spine	TTM	tensor tympani muscle
MAC	mastoid air cells	ZP	zygomatic process

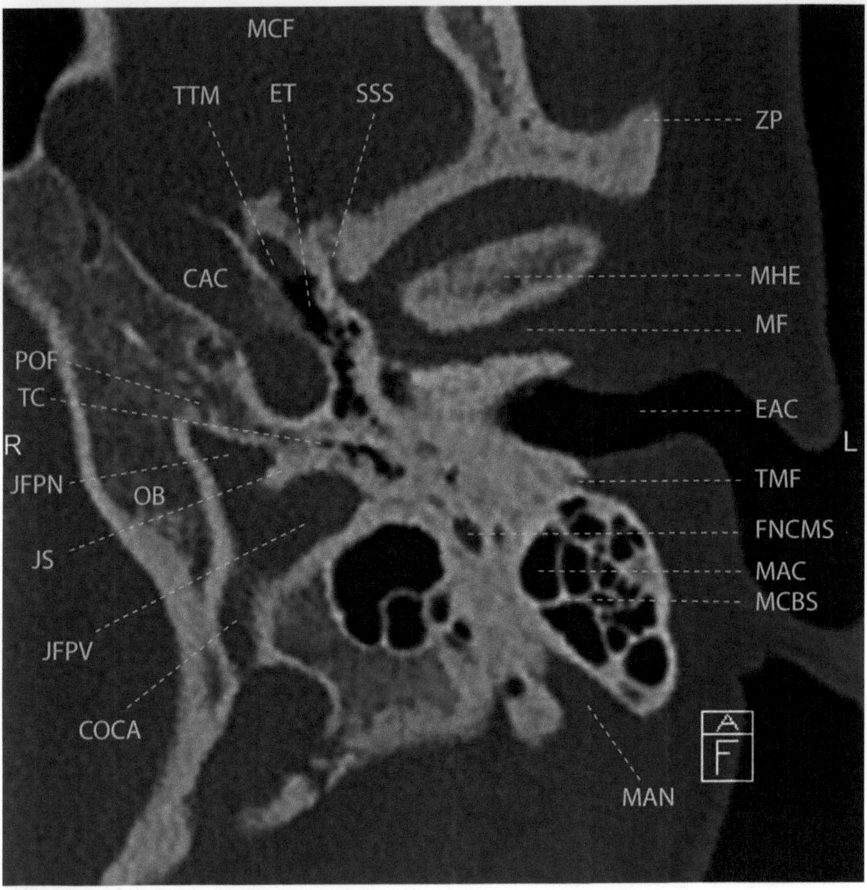

A41: HRCT axial reformat

CAC	carotid canal	MAN	mastoid notch
COCA	condylar canal	MCBS	mastoid cells bony septa
EAC	external auditory canal	MCF	middle cranial fossa
ET	Eustachian tube	MF	mandibular fossa
FNCMS	facial nerve canal – mastoid segment	MHE	mandible – head
JFPN	jugular foramen – pars nervosa	OB	occipital bone
JFPV	jugular foramen – pars vascularis	POF	petrooccipital fissure
JS	jugular spine	SSS	sphenosquamosal suture
MAC	mastoid air cells	TC	tympanic canaliculus
		TMF	tympanomastoid fissure
		TTM	tensor tympani muscle
		ZP	zygomatic process

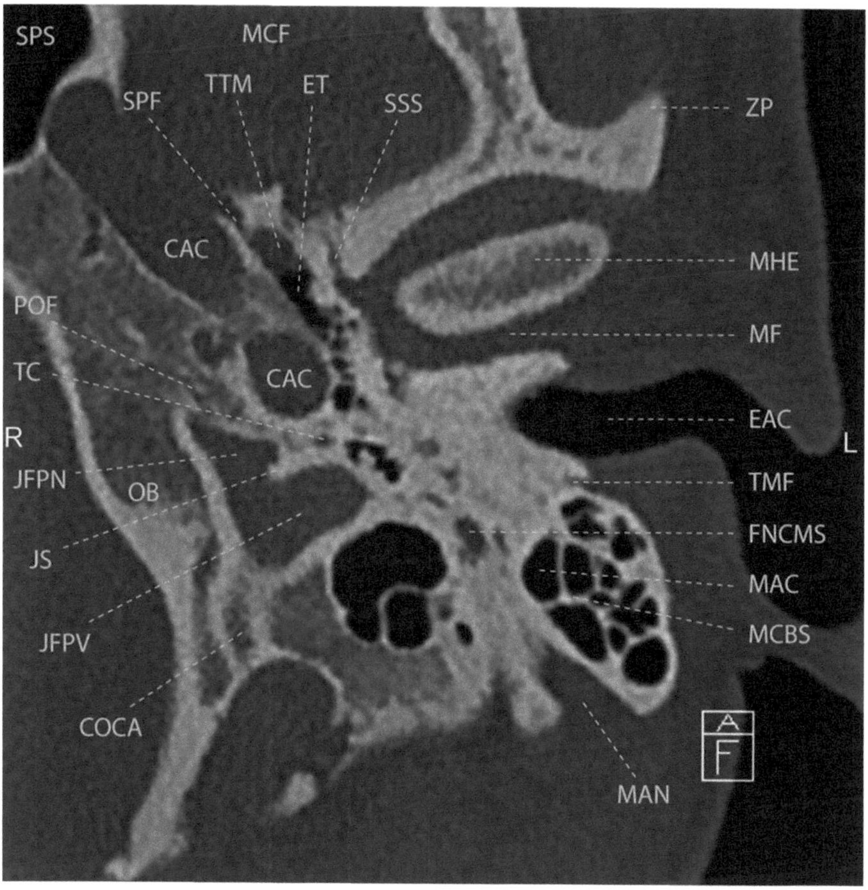

A42: HRCT axial reformat

CAC	carotid canal	MCBS	mastoid cells bony septa
COCA	condylar canal	MCF	middle cranial fossa
EAC	external auditory canal	MF	mandibular fossa
ET	Eustachian tube	MHE	mandible – head
FNCMS	facial nerve canal – mastoid segment	OB	occipital bone
JFPN	jugular foramen – pars nervosa	POF	petrooccipital fissure
JFPV	jugular foramen – pars vascularis	SPF	sphenopetrosal fissure
JS	jugular spine	SPS	sphenoidal sinus
MAC	mastoid air cells	SSS	sphenosquamosal suture
MAN	mastoid notch	TC	tympanic canaliculus
		TMF	tympanomastoid fissure
		TTM	tensor tympani muscle
		ZP	zygomatic process

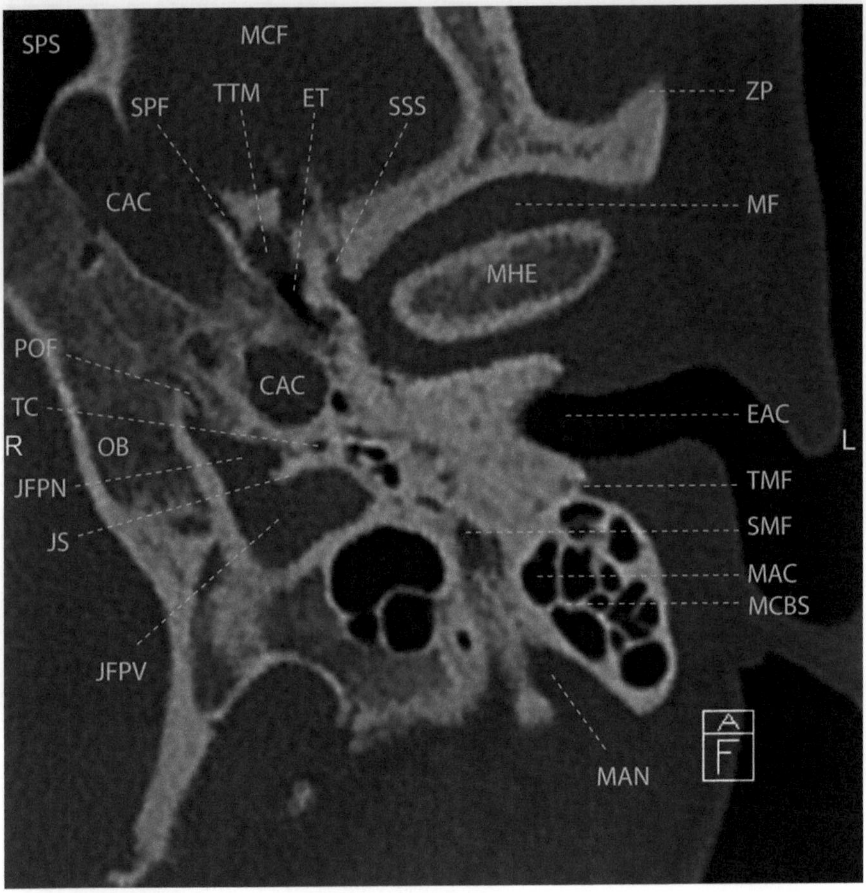

A43: HRCT axial reformat

CAC	carotid canal
EAC	external auditory canal
ET	Eustachian tube
JFPN	jugular foramen – pars nervosa
JFPV	jugular foramen – pars vascularis
JS	jugular spine
MAC	mastoid air cells
MAN	mastoid notch
MCBS	mastoid cells bony septa
MCF	middle cranial fossa
MF	mandibular fossa
MHE	mandible – head
OB	occipital bone
POF	petrooccipital fissure
SMF	stylomastoid foramen
SPF	sphenopetrosal fissure
SPS	sphenoidal sinus
SSS	sphenosquamosal suture
TC	tympanic canaliculus
TMF	tympanomastoid fissure
TTM	tensor tympani muscle
ZP	zygomatic process

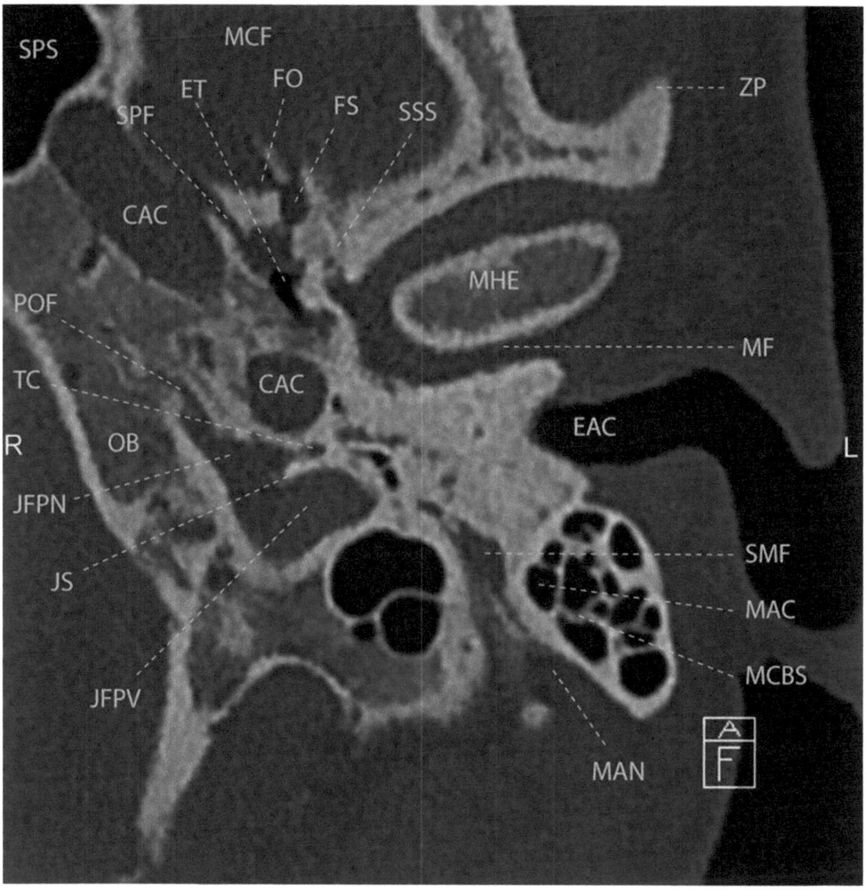

A44: HRCT axial reformat

CAC	carotid canal	MCBS	mastoid cells bony septa
EAC	external auditory canal	MCF	middle cranial fossa
ET	Eustachian tube	MF	mandibular fossa
FO	foramen ovale	MHE	mandible – head
FS	foramen spinosum	OB	occipital bone
JFPN	jugular foramen – pars nervosa	POF	petrooccipital fissure
		SMF	stylomastoid foramen
JFPV	jugular foramen – pars vascularis	SPF	sphenopetrosal fissure
		SPS	sphenoidal sinus
JS	jugular spine	SSS	sphenosquamosal suture
MAC	mastoid air cells	TC	tympanic canaliculus
MAN	mastoid notch	ZP	zygomatic process

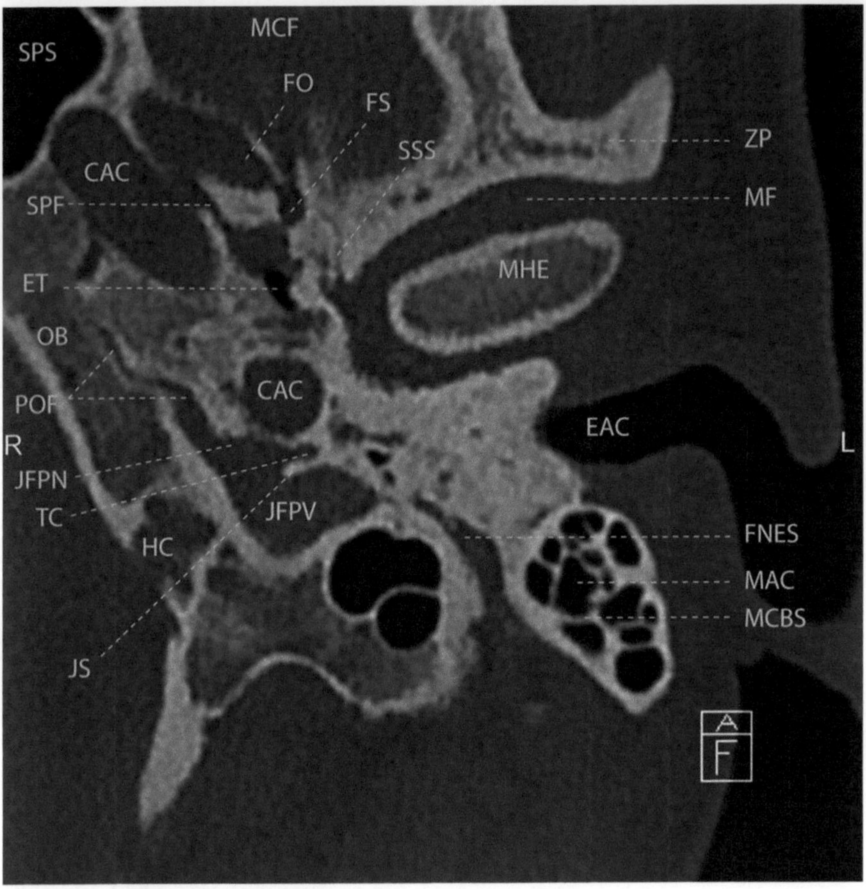

A45: HRCT axial reformat

CAC carotid canal
EAC external auditory canal
ET Eustachian tube
FNES facial nerve – extracranial segment
FO foramen ovale
FS foramen spinosum
HC hypoglossal canal
JFPN jugular foramen – pars nervosa
JFPV jugular foramen – pars vascularis
JS jugular spine
MAC mastoid air cells
MCBS mastoid cells bony septa
MCF middle cranial fossa
MF mandibular fossa
MHE mandible – head
OB occipital bone
POF petrooccipital fissure
SPF sphenopetrosal fissure
SPS sphenoidal sinus
SSS sphenosquamosal suture
TC tympanic canaliculus
ZP zygomatic process

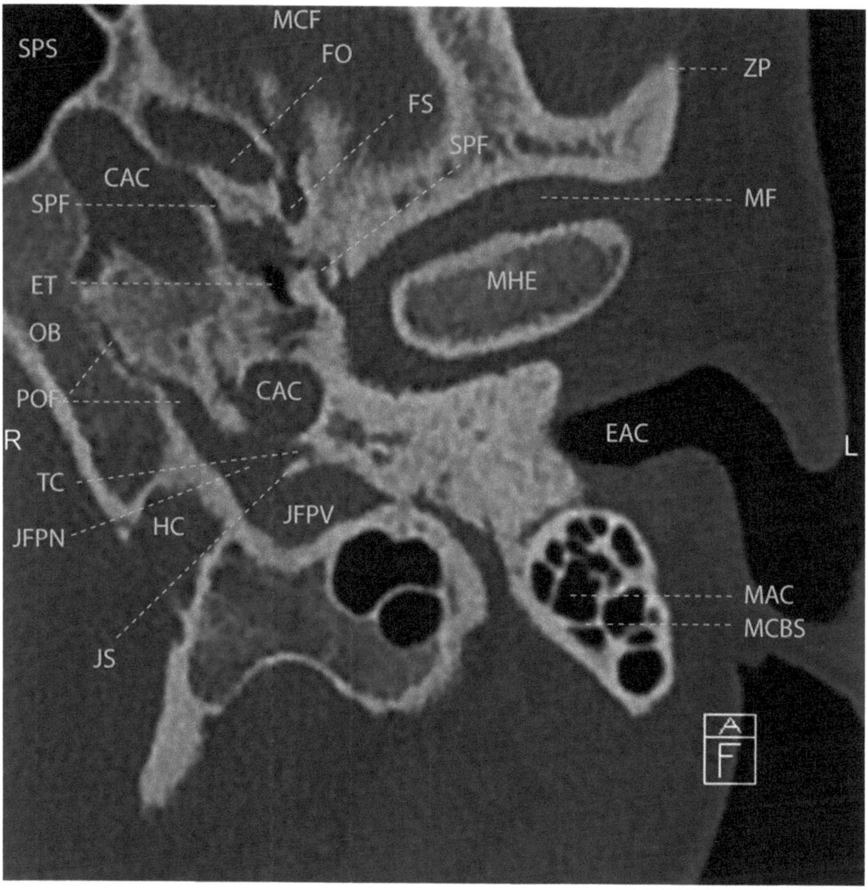

A46: HRCT axial reformat

CAC	carotid canal	MAC	mastoid air cells
EAC	external auditory canal	MCBS	mastoid cells bony septa
ET	Eustachian tube	MCF	middle cranial fossa
FO	foramen ovale	MF	mandibular fossa
FS	foramen spinosum	MHE	mandible – head
HC	hypoglossal canal	OB	occipital bone
JFPN	jugular foramen – pars nervosa	POF	petrooccipital fissure
		SPF	sphenopetrosal fissure
JFPV	jugular foramen – pars vascularis	SPS	sphenoidal sinus
		TC	tympanic canaliculus
JS	jugular spine	ZP	zygomatic process

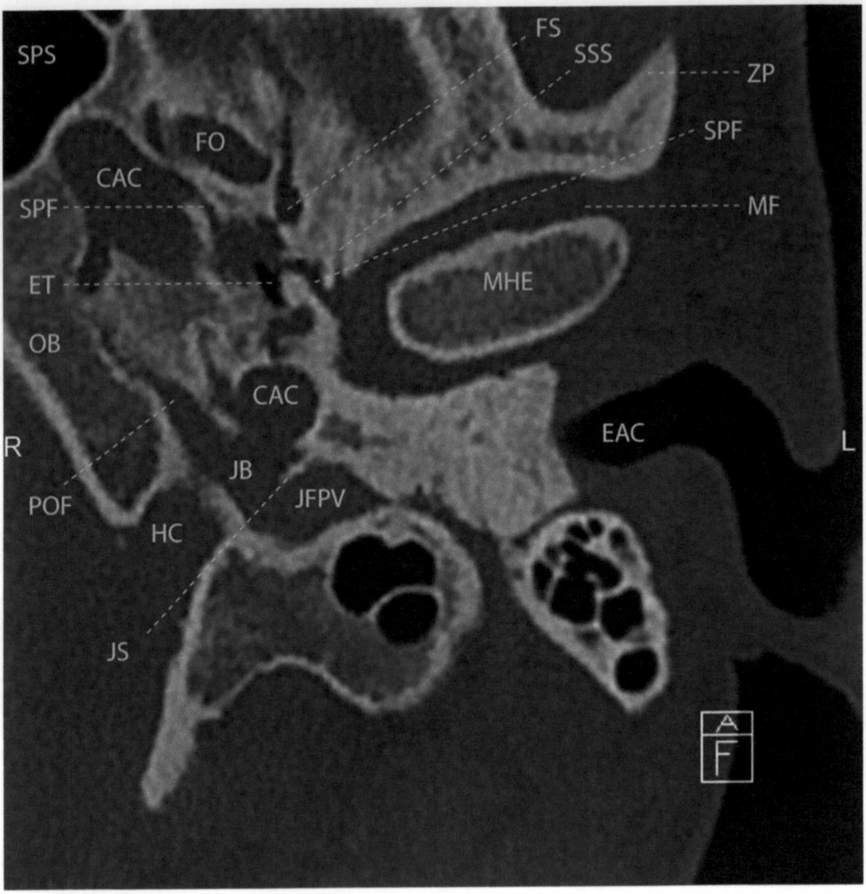

A47: HRCT axial reformat

CAC	carotid canal
EAC	external auditory canal
ET	Eustachian tube
FO	foramen ovale
FS	foramen spinosum
HC	hypoglossal canal
JB	jugular bulb
JFPV	jugular foramen – pars vascularis
JS	jugular spine
MF	mandibular fossa
MHE	mandible – head
OB	occipital bone
POF	petrooccipital fissure
SPF	sphenopetrosal fissure
SPS	sphenoidal sinus
SSS	sphenosquamosal suture
ZP	zygomatic process

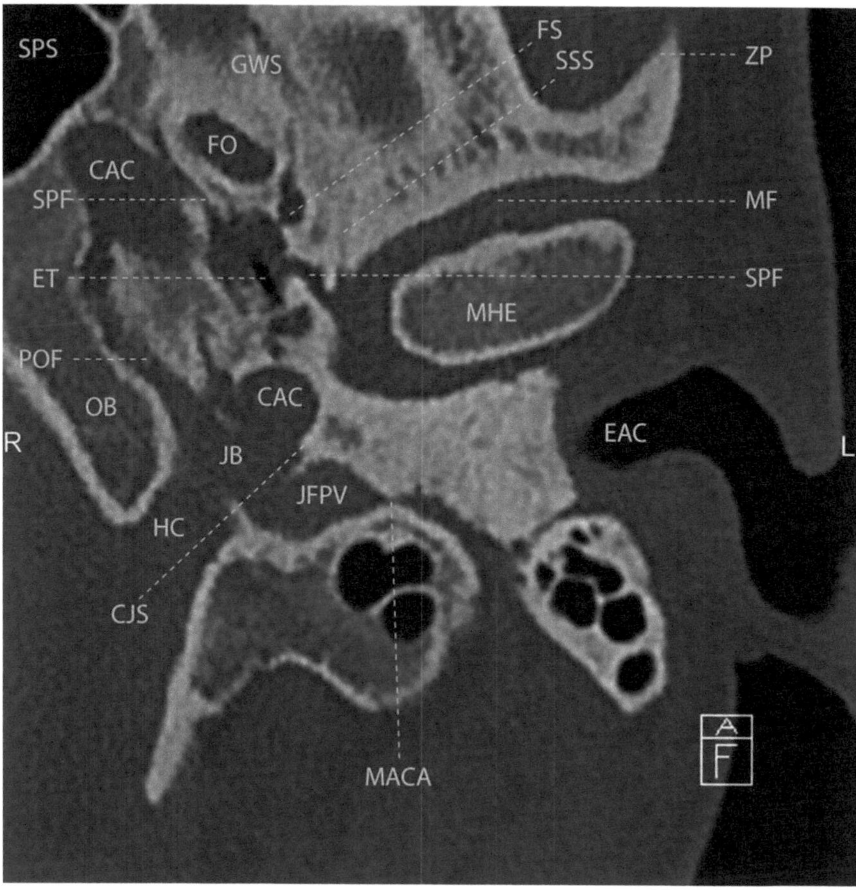

A48: HRCT axial reformat

CAC	carotid canal	MACA	mastoid canaliculus
CJS	caroticojugular spine	MF	mandibular fossa
EAC	external auditory canal	MHE	mandible – head
ET	Eustachian tube	OB	occipital bone
FO	foramen ovale	POF	petrooccipital fissure
FS	foramen spinosum	SPF	sphenopetrosal fissure
GWS	greater wing of sphenoid	SPS	sphenoidal sinus
HC	hypoglossal canal	SSS	sphenosquamosal suture
JB	jugular bulb	ZP	zygomatic process
JFPV	jugular foramen pars – vascularis		

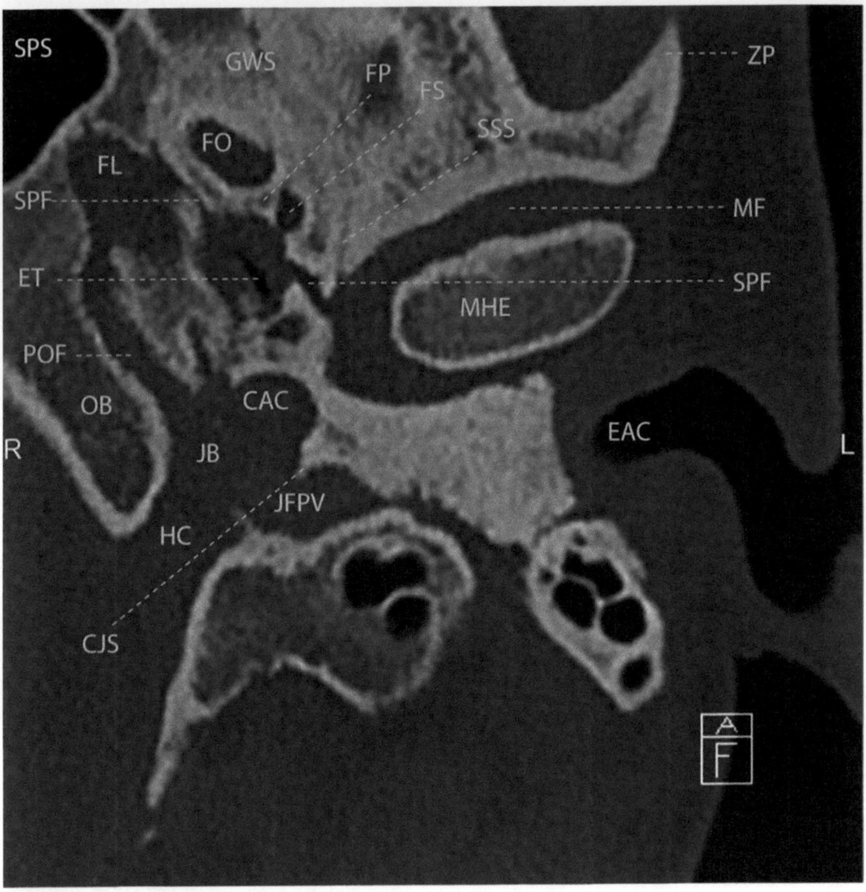

A49: HRCT axial reformat

CAC	carotid canal
CJS	caroticojugular spine
EAC	external auditory canal
ET	Eustachian tube
FL	foramen lacerum
FO	foramen ovale
FP	foramen petrosum
FS	foramen spinosum
GWS	greater wing of sphenoid
HC	hypoglossal canal
JB	jugular bulb
JFPV	jugular foramen – pars vascularis
MF	mandibular fossa
MHE	mandible – head
OB	occipital bone
POF	petrooccipital fissure
SPF	sphenopetrosal fissure
SPS	sphenoidal sinus
SSS	sphenosquamosal suture
ZP	zygomatic process

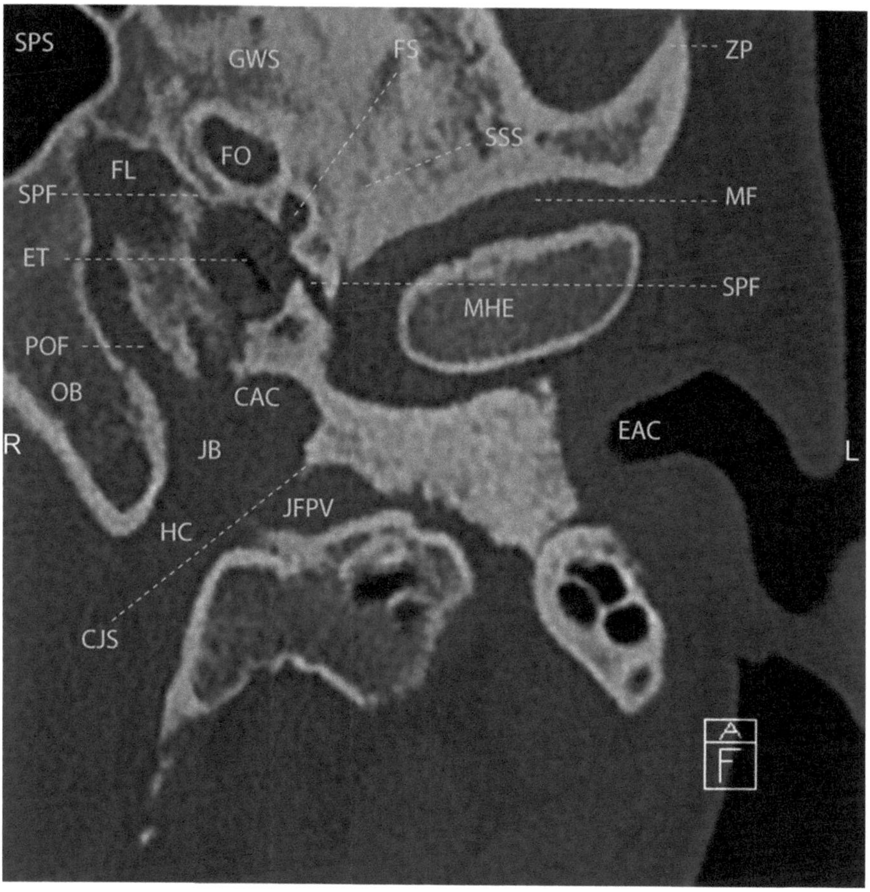

A50: HRCT axial reformat

CAC	carotid canal	CJS	caroticojugular spine
EAC	external auditory canal	MF	mandibular fossa
ET	Eustachian tube	MHE	mandible – head
FL	foramen lacerum	OB	occipital bone
FO	foramen ovale	POF	petrooccipital fissure
FS	foramen spinosum	SPF	sphenopetrosal fissure
GWS	greater wing of sphenoid	SPS	sphenoidal sinus
HC	hypoglossal canal	SSS	sphenosquamosal suture
JB	jugular bulb	ZP	zygomatic process
JFPV	jugular foramen – pars vascularis		

Chapter 3
HRCT Coronal Reformat

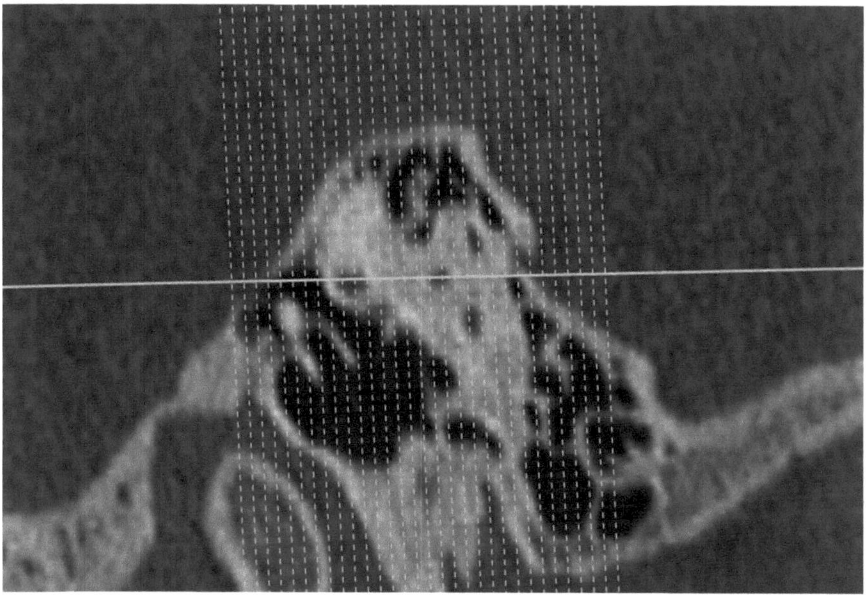

C00: HRCT coronal reformat

Coronal plane

Using an MPR 3D viewer, anterior and posterior aspects of the lateral semicircular canal (which define the axial plane) are identified. The coronal plane is perpendicular to the axial plane.

J. Kopřiva, J. Žižka, *Temporal Bone CT and MRI Anatomy: A Guide to 3D Volumetric Acquisitions*, DOI 10.1007/978-3-319-08242-4_3

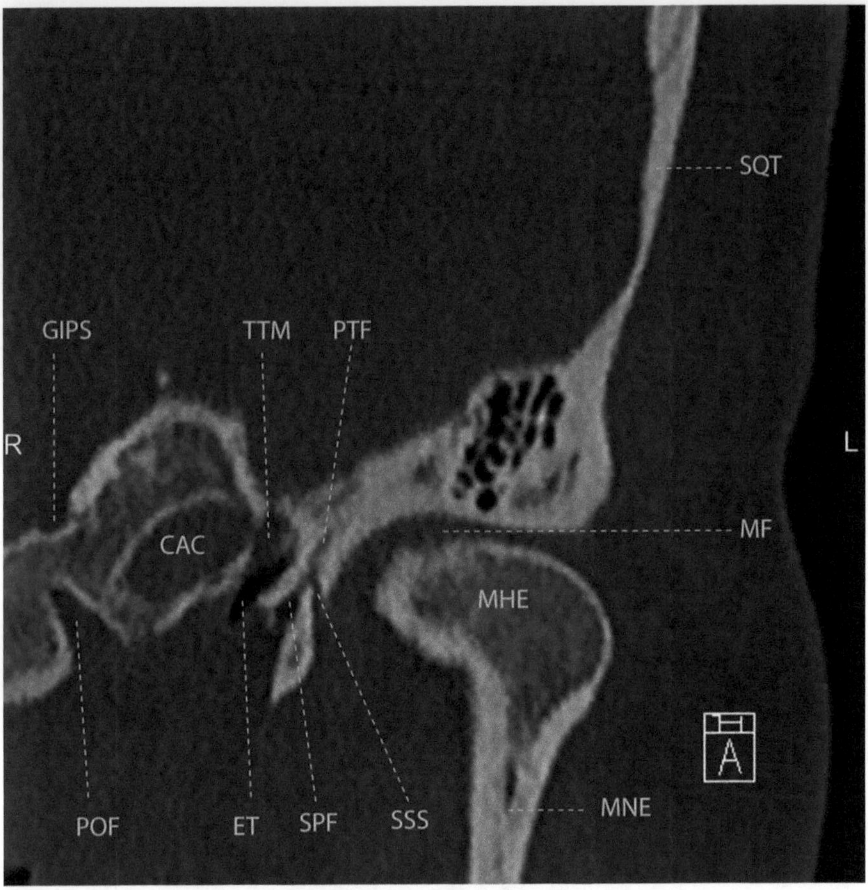

C01: HRCT coronal reformat

CAC	carotid canal
ET	Eustachian tube
GIPS	groove of the inferior petrosal sinus
MF	mandibular fossa
MHE	mandible – head
MNE	mandible – neck
POF	petrooccipital fissure
PTF	petrotympanic fissure
SPF	sphenopetrosal fissure
SQT	squama temporalis
SSS	sphenosquamosal suture
TTM	tensor tympani muscle

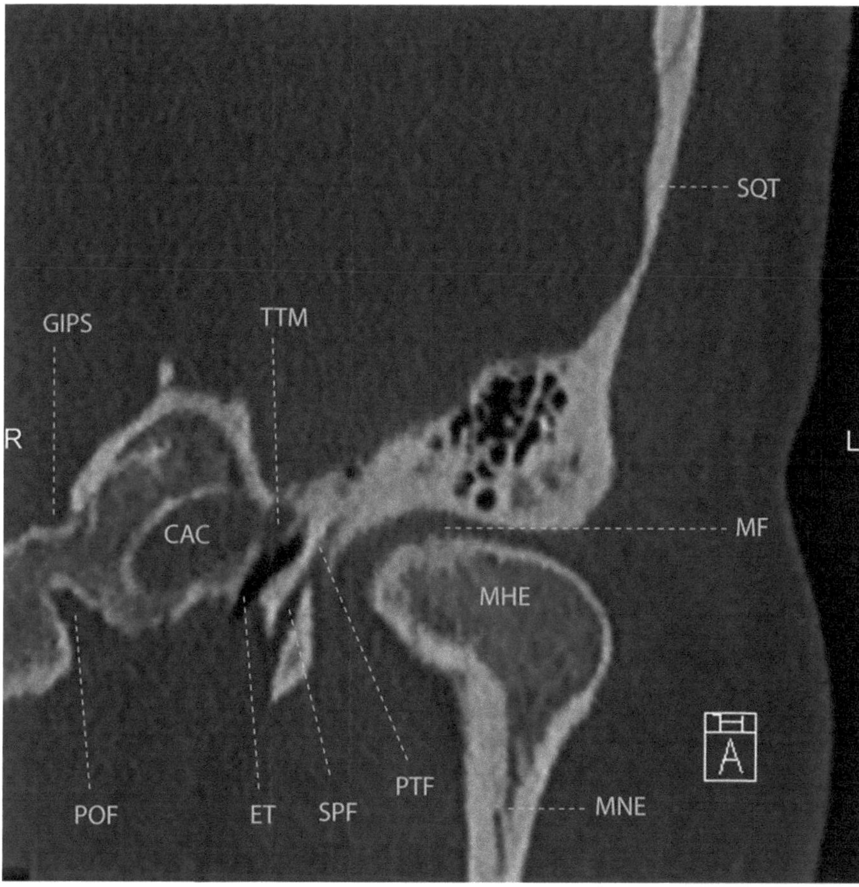

C02: HRCT coronal reformat

CAC	carotid canal	MNE	mandible – neck
ET	Eustachian tube	POF	petrooccipital fissure
GIPS	groove of the inferior petrosal sinus	PTF	petrotympanic fissure
		SPF	sphenopetrosal fissure
MF	mandibular fossa	SQT	squama temporalis
MHE	mandible – head	TTM	tensor tympani muscle

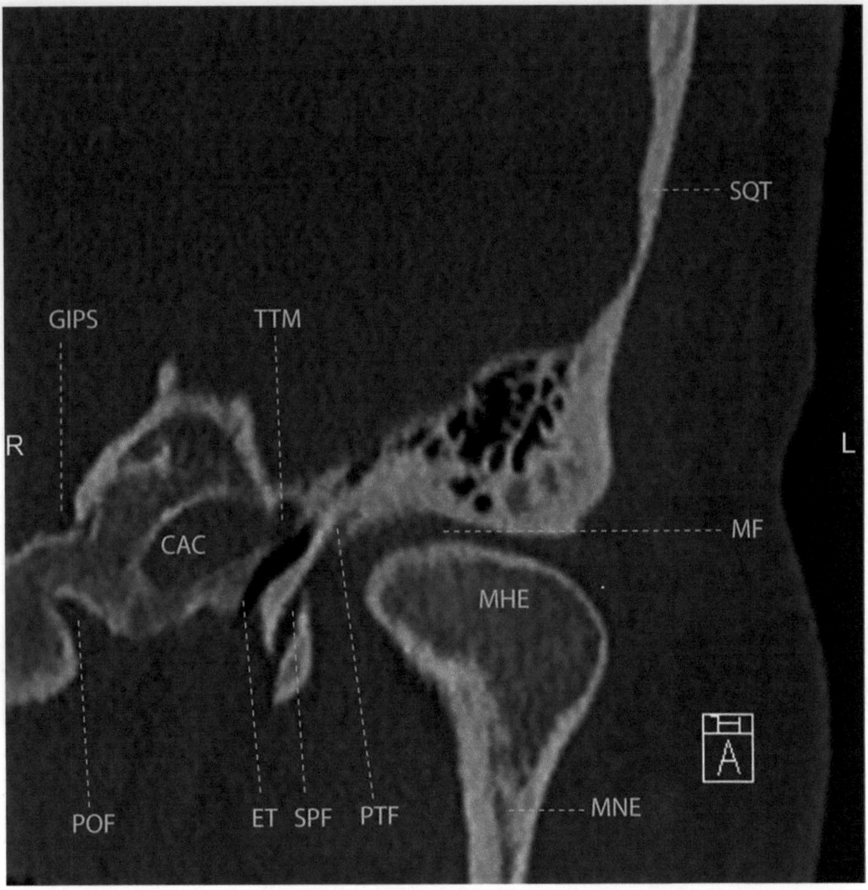

C03: HRCT coronal reformat

CAC	carotid canal	MNE	mandible – neck
ET	Eustachian tube	POF	petrooccipital fissure
GIPS	groove of the inferior petrosal sinus	PTF	petrotympanic fissure
MF	mandibular fossa	SPF	sphenopetrosal fissure
MHE	mandible – head	SQT	squama temporalis
		TTM	tensor tympani muscle

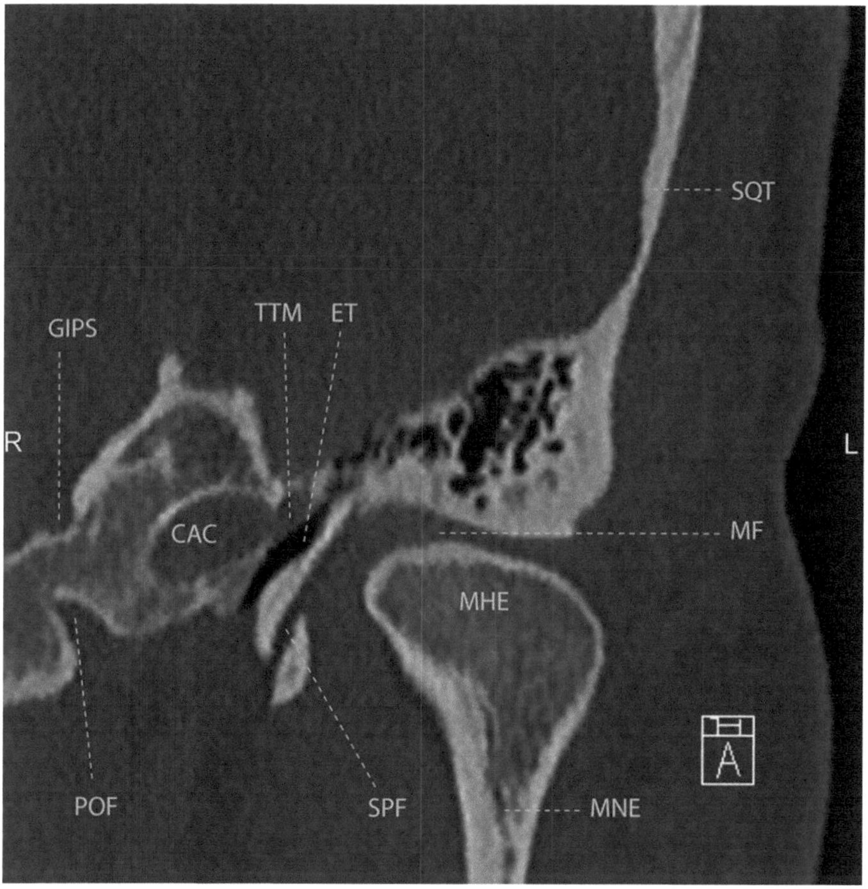

C04: HRCT coronal reformat

CAC	carotid canal	MNE	mandible – neck
ET	Eustachian tube	POF	petrooccipital fissure
GIPS	groove of the inferior petrosal sinus	SPF	sphenopetrosal fissure
MF	mandibular fossa	SQT	squama temporalis
MHE	mandible – head	TTM	tensor tympani muscle

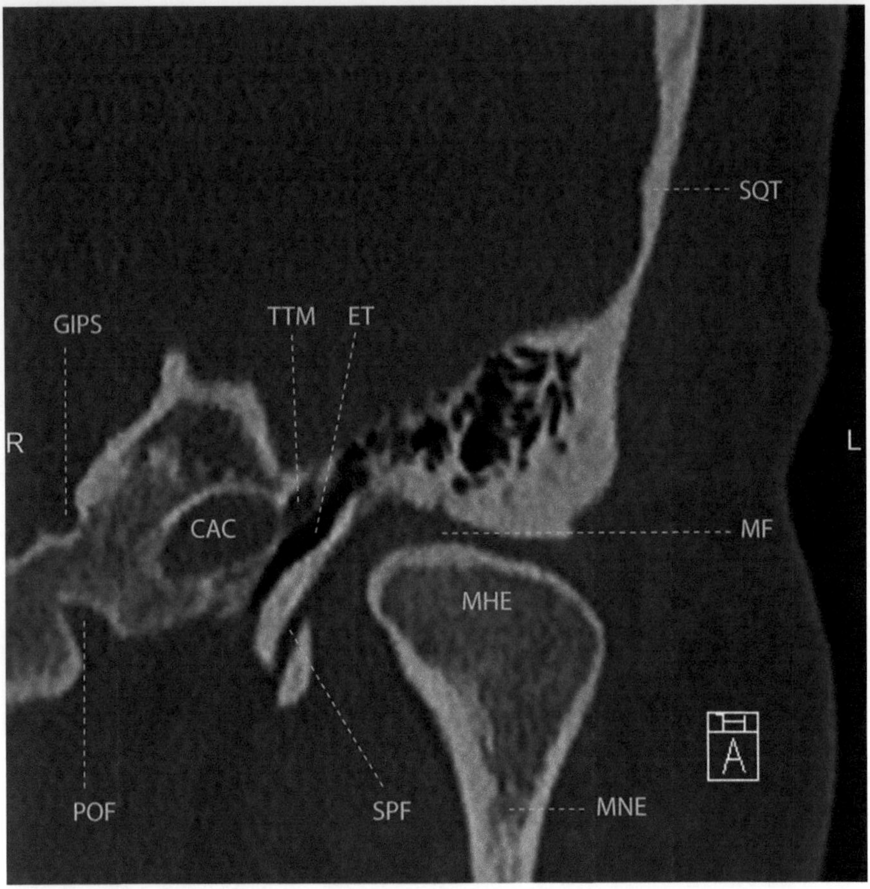

C05: HRCT coronal reformat

CAC carotid canal
ET Eustachian tube
GIPS groove of the inferior petrosal sinus
MF mandibular fossa
MHE mandible – head
MNE mandible – neck
POF petrooccipital fissure
SPF sphenopetrosal fissure
SQT squama temporalis
TTM tensor tympani muscle

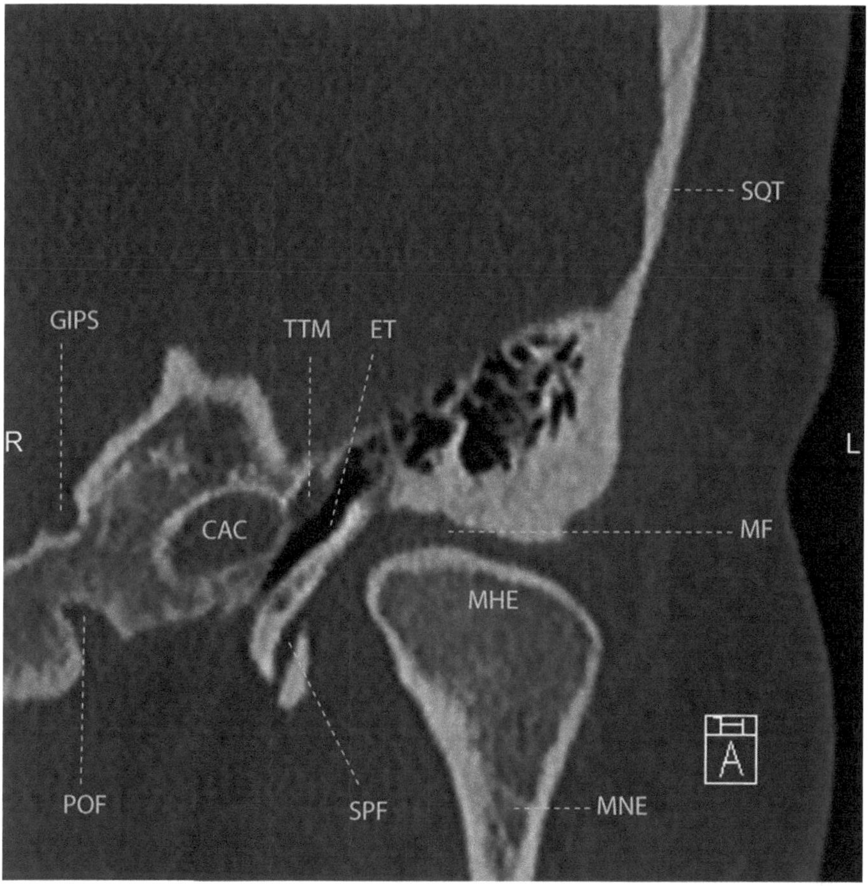

C06: HRCT coronal reformat

CAC	carotid canal
ET	Eustachian tube
GIPS	groove of the inferior petrosal sinus
MF	mandibular fossa
MHE	mandible – head
MNE	mandible – neck
POF	petrooccipital fissure
SPF	sphenopetrosal fissure
SQT	squama temporalis
TTM	tensor tympani muscle

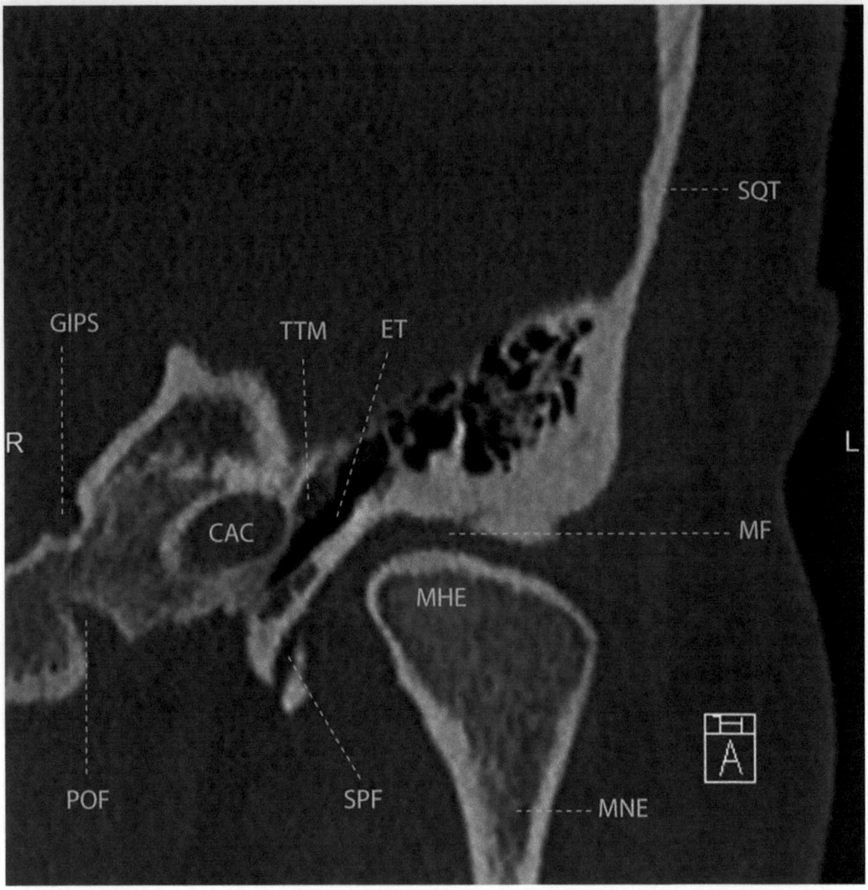

C07: HRCT coronal reformat

CAC	carotid canal
ET	Eustachian tube
GIPS	groove of the inferior petrosal sinus
MF	mandibular fossa
MHE	mandible – head
MNE	mandible – neck
POF	petrooccipital fissure
SPF	sphenopetrosal fissure
SQT	squama temporalis
TTM	tensor tympani muscle

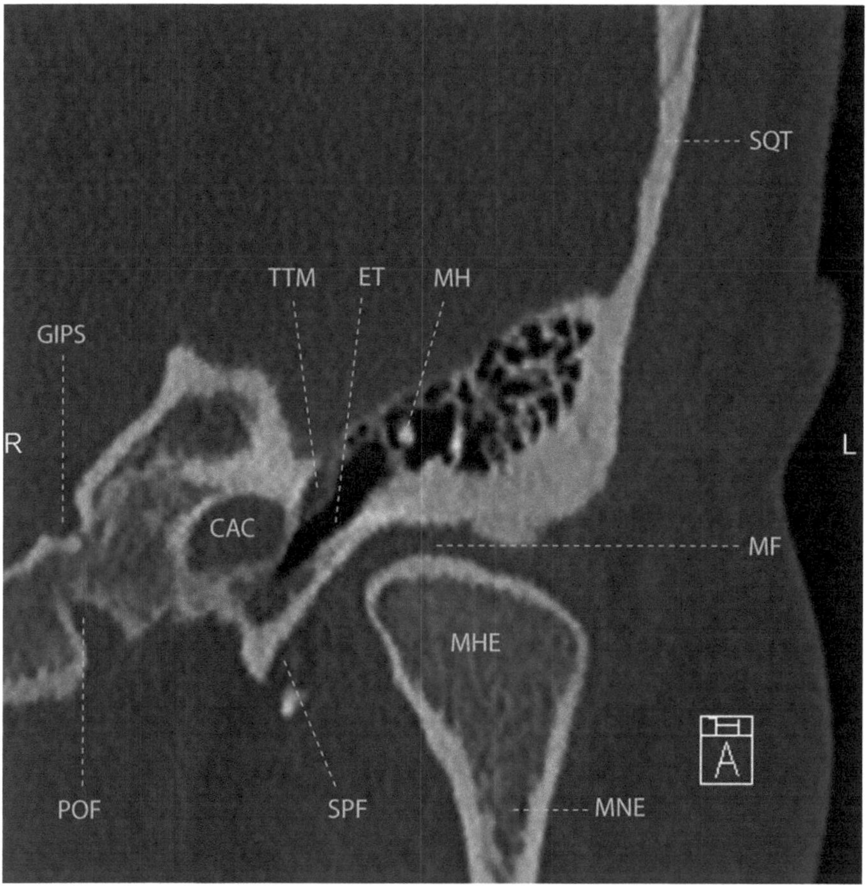

C08: HRCT coronal reformat

CAC carotid canal
ET Eustachian tube
GIPS groove of the inferior petrosal sinus
MF mandibular fossa
MH malleus – head
MHE mandible – head
MNE mandible – neck
POF petrooccipital fissure
SPF sphenopetrosal fissure
SQT squama temporalis
TTM tensor tympani muscle

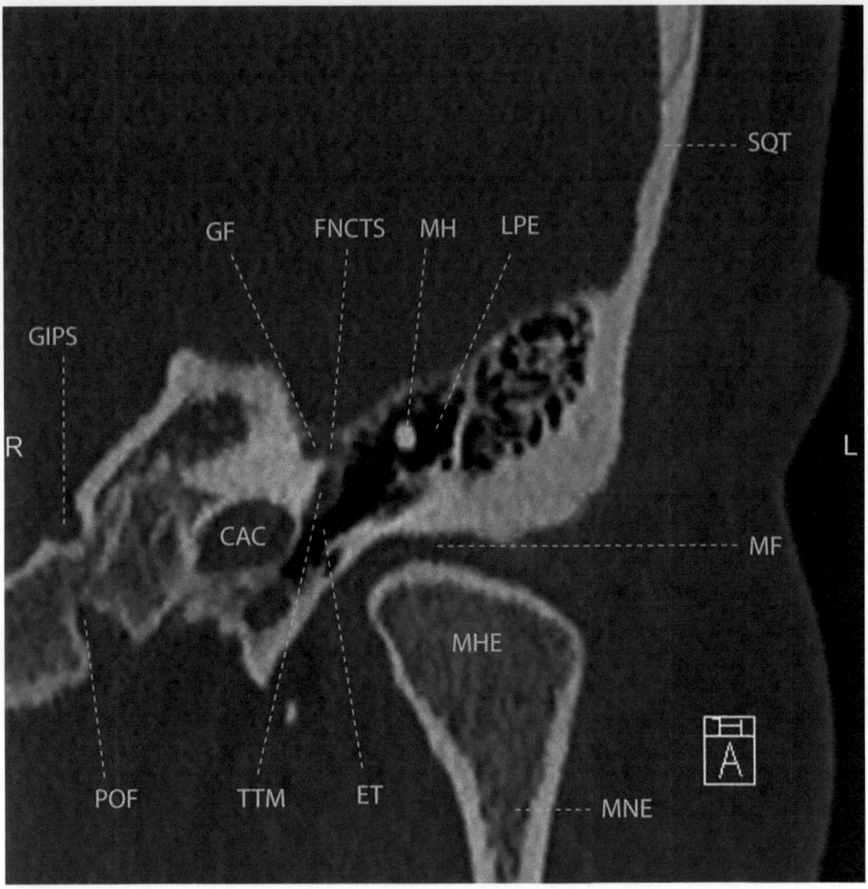

C09: HRCT coronal reformat

CAC	carotid canal	MF	mandibular fossa
ET	Eustachian tube	MH	malleus – head
FNCTS	facial nerve canal – tympanic segment	MHE	mandible – head
GF	geniculate fossa	MNE	mandible – neck
GIPS	groove of the inferior petrosal sinus	POF	petrooccipital fissure
LPE	lateral posterior epitympanum	SQT	squama temporalis
		TTM	tensor tympani muscle

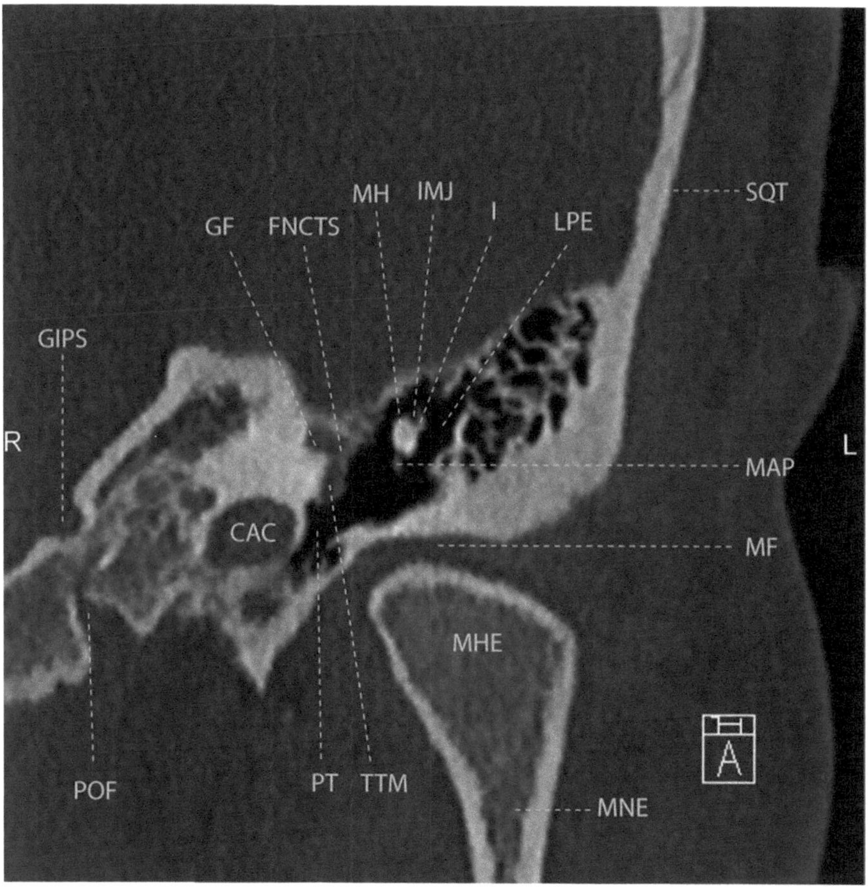

C10: HRCT coronal reformat

CAC	carotid canal
FNCTS	facial nerve canal – tympanic segment
GF	geniculate fossa
GIPS	groove of the inferior petrosal sinus
I	incus
IMJ	incudomallear joint
LPE	lateral posterior epitympanum
MAP	malleus – anterior process
MF	mandibular fossa
MH	malleus – head
MHE	mandible – head
MNE	mandible – neck
POF	petrooccipital fissure
PT	protympanum
SQT	squama temporalis
TTM	tensor tympani muscle

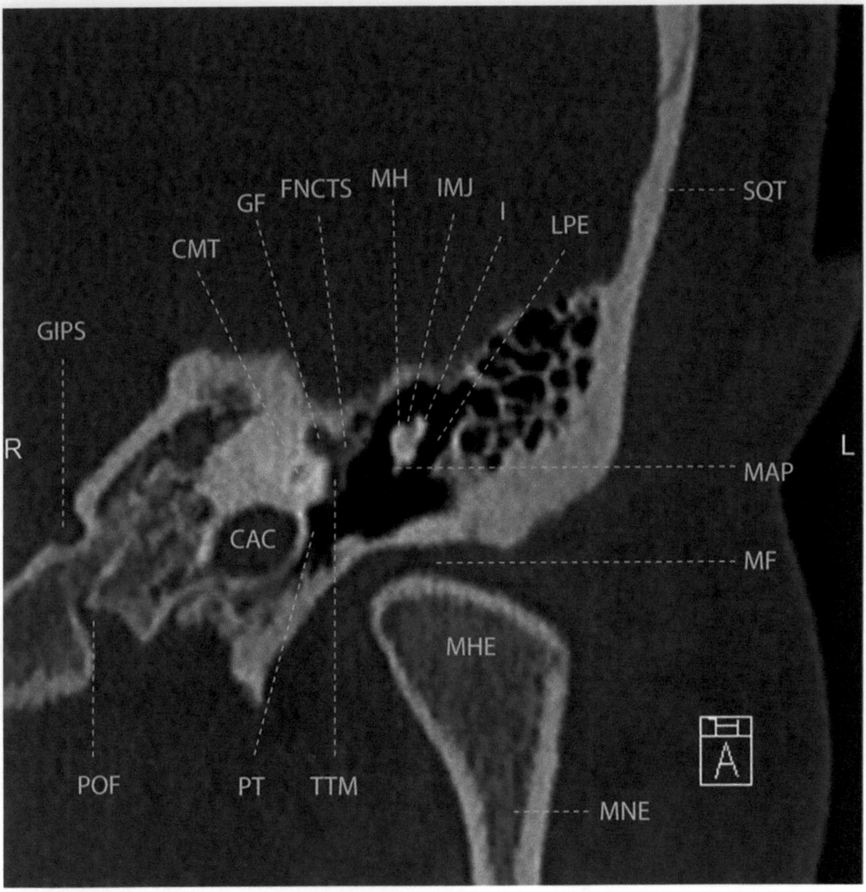

C11: HRCT coronal reformat

CAC	carotid canal
CMT	cochlea – middle turn
FNCTS	facial nerve canal – tympanic segment
GF	geniculate fossa
GIPS	groove of the inferior petrosal sinus
I	incus
IMJ	incudomallear joint
LPE	lateral posterior epitympanum
MAP	malleus – anterior process
MF	mandibular fossa
MH	malleus – head
MHE	mandible – head
MNE	mandible – neck
POF	petrooccipital fissure
PT	protympanum
SQT	squama temporalis
TTM	tensor tympani muscle

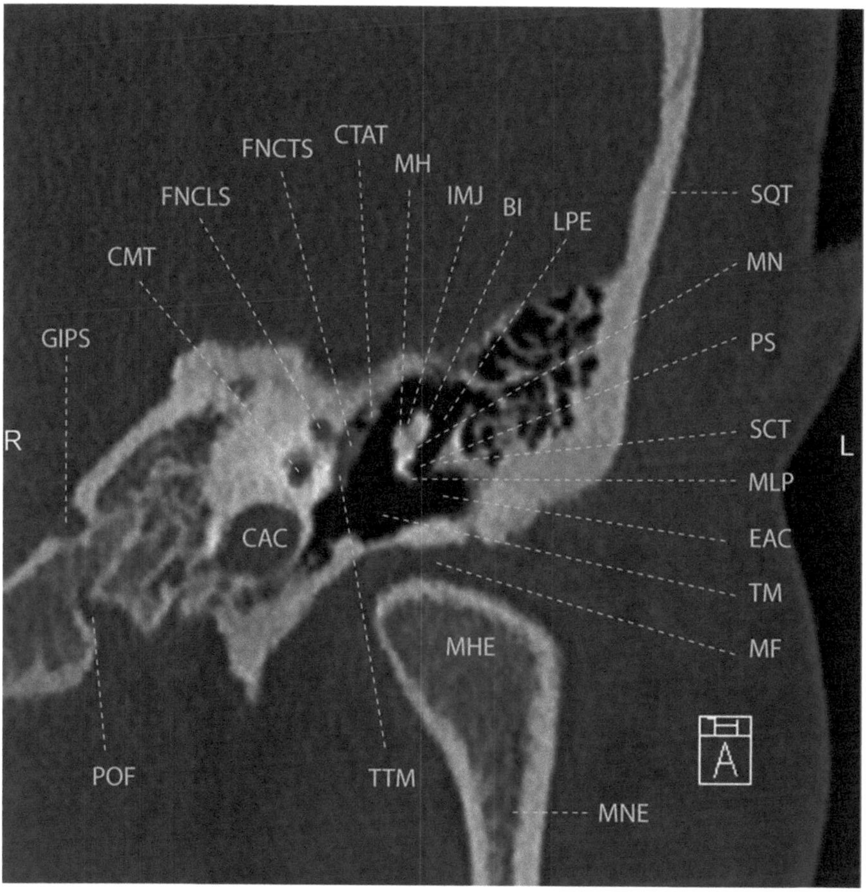

C12: HRCT coronal reformat

BI	body of incus
CAC	carotid canal
CMT	cochlea – middle turn
CTAT	crista transversa anterior epitympanum
EAC	external auditory canal
FNCLS	facial nerve canal – labyrinthine segment
FNCTS	facial nerve canal – tympanic segment
GIPS	groove of the inferior petrosal sinus
IMJ	incudomallear joint
LPE	lateral posterior epitympanum
MF	mandibular fossa
MH	malleus – head
MHE	mandible – head
MLP	malleus – lateral process
MN	malleus – neck
MNE	mandible neck
POF	petrooccipital fissure
PS	Prussak's space
SCT	scutum
SQT	squama temporalis
TM	tympanic membrane
TTM	tensor tympani muscle

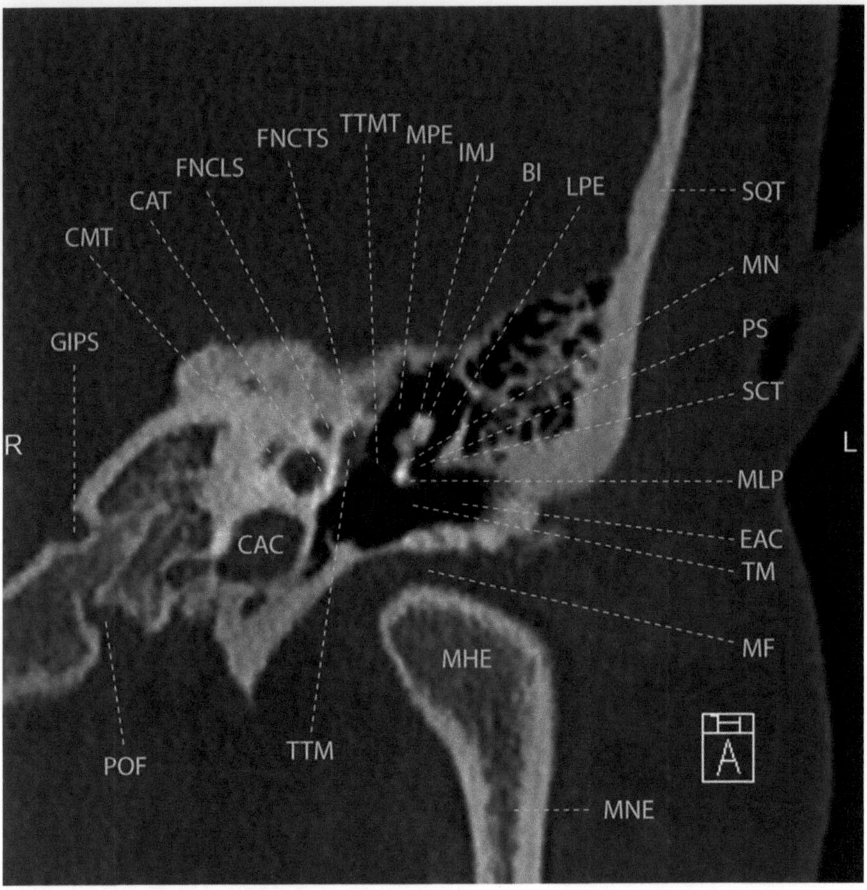

C13: HRCT coronal reformat

BI body of incus
CAC carotid canal
CAT cochlea – apical turn
CMT cochlea – middle turn
EAC external auditory canal
FNCLS facial nerve canal – labyrinthine segment
FNCTS facial nerve canal – tympanic segment
GIPS groove of the inferior petrosal sinus
IMJ incudomallear joint
LPE lateral posterior epitympanum
MF mandibular fossa
MHE mandible – head
MLP malleus – lateral process
MN malleus – neck
MNE mandible – neck
MPE medial posterior epitympanum
POF petrooccipital fissure
PS Prussak's space
SCT scutum
SQT squama temporalis
TM tympanic membrane
TTM tensor tympani muscle
TTMT tensor tympani muscle – tendon

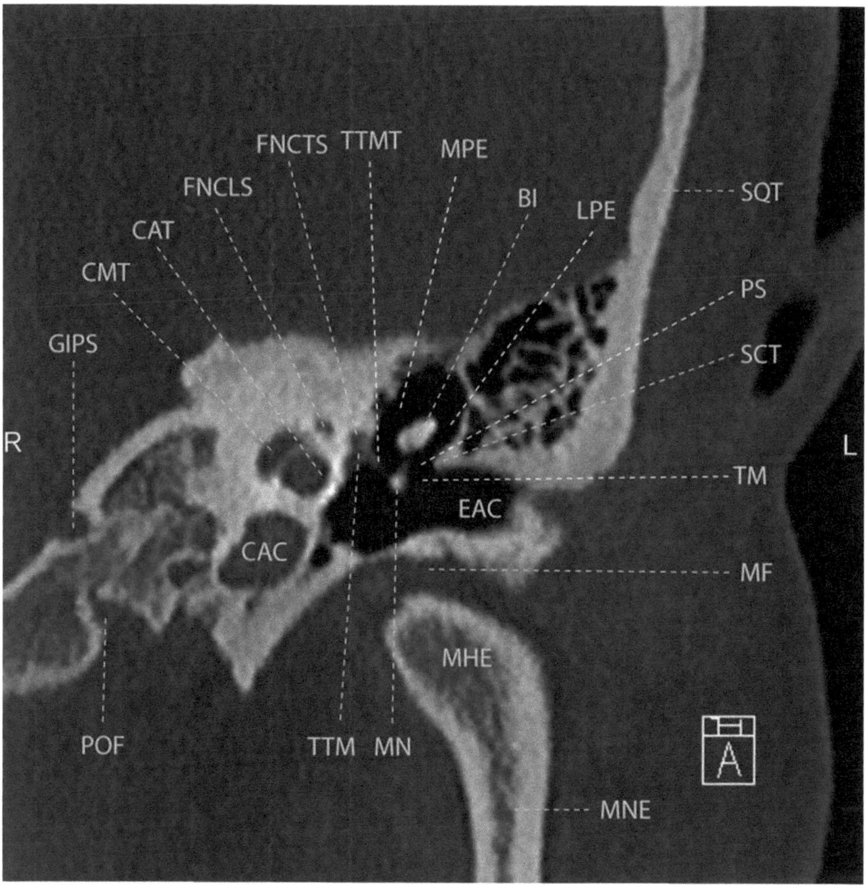

C14: HRCT coronal reformat

BI	body of incus
CAC	carotid canal
CAT	cochlea – apical turn
CMT	cochlea – middle turn
EAC	external auditory canal
FNCLS	facial nerve canal – labyrinthine segment
FNCTS	facial nerve canal – tympanic segment
GIPS	groove of the inferior petrosal sinus
LPE	lateral posterior epitympanum
MF	mandibular fossa
MHE	mandible – head
MN	malleus – neck
MNE	mandible neck
MPE	medial posterior epitympanum
POF	petrooccipital fissure
PS	Prussak's space
SCT	scutum
SQT	squama temporalis
TM	tympanic membrane
TTM	tensor tympani muscle
TTMT	tensor tympani muscle – tendon

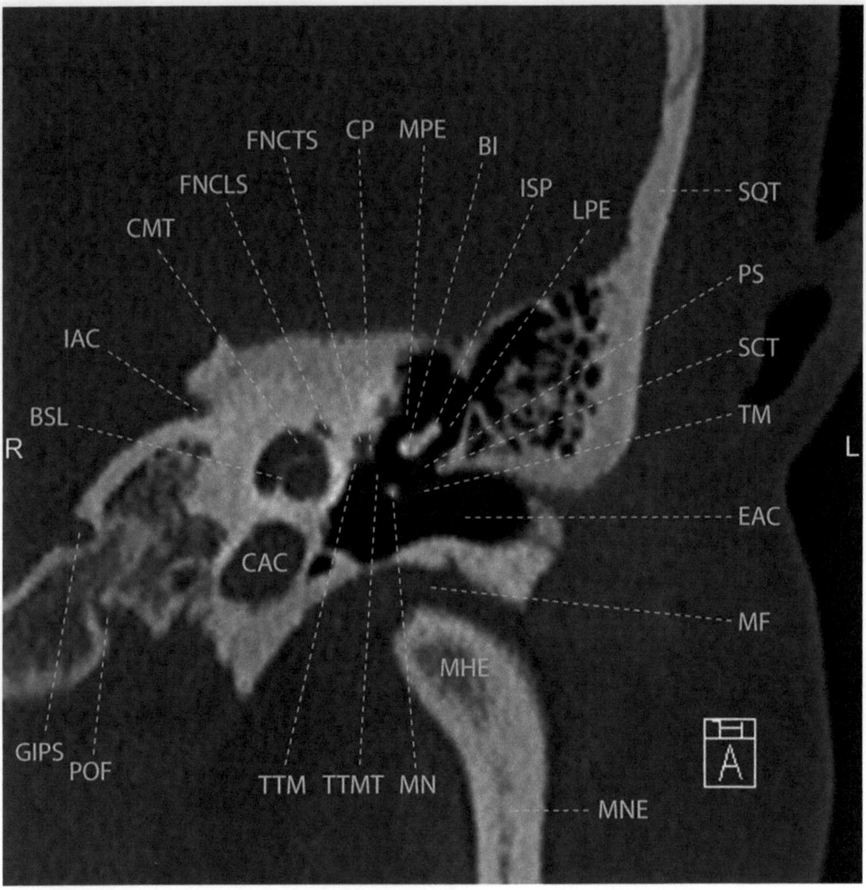

C15: HRCT coronal reformat

BI	body of incus
BSL	bony spiral lamina
CAC	carotid canal
CMT	cochlea – middle turn
CP	cochleariform process
EAC	external auditory canal
FNCLS	facial nerve canal – labyrinthine segment
FNCTS	facial nerve canal – tympanic segment
GIPS	groove of the inferior petrosal sinus
IAC	internal auditory canal
ISP	incus – short process
LPE	lateral posterior epitympanum
MF	mandibular fossa
MHE	mandible – head
MN	malleus – neck
MNE	mandible – neck
MPE	medial posterior epitympanum
POF	petrooccipital fissure
PS	Prussak's space
SCT	scutum
SQT	squama temporalis
TM	tympanic membrane
TTM	tensor tympani muscle
TTMT	tensor tympani muscle – tendon

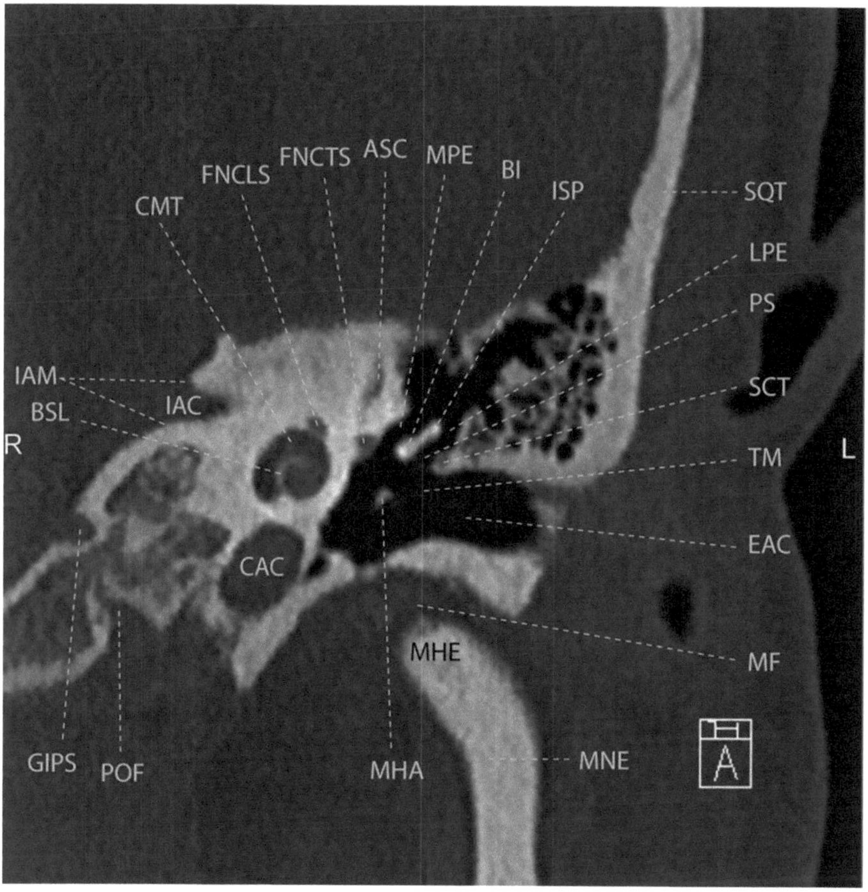

C16: HRCT coronal reformat

ASC	anterior semicircular canal	IAM	internal auditory canal – meatus
BI	body of incus	ISP	incus – short process
BSL	bony spiral lamina	LPE	lateral posterior epitympanum
CAC	carotid canal	MF	mandibular fossa
CMT	cochlea – middle turn	MHA	malleus – handle
EAC	external auditory canal	MHE	mandible – head
FNCLS	facial nerve canal – labyrinthine segment	MNE	mandible – neck
		MPE	medial posterior epitympanum
FNCTS	facial nerve canal – tympanic segment	POF	petrooccipital fissure
		PS	Prussak's space
GIPS	groove of the inferior petrosal sinus	SCT	scutum
		SQT	squama temporalis
IAC	internal auditory canal	TM	tympanic membrane

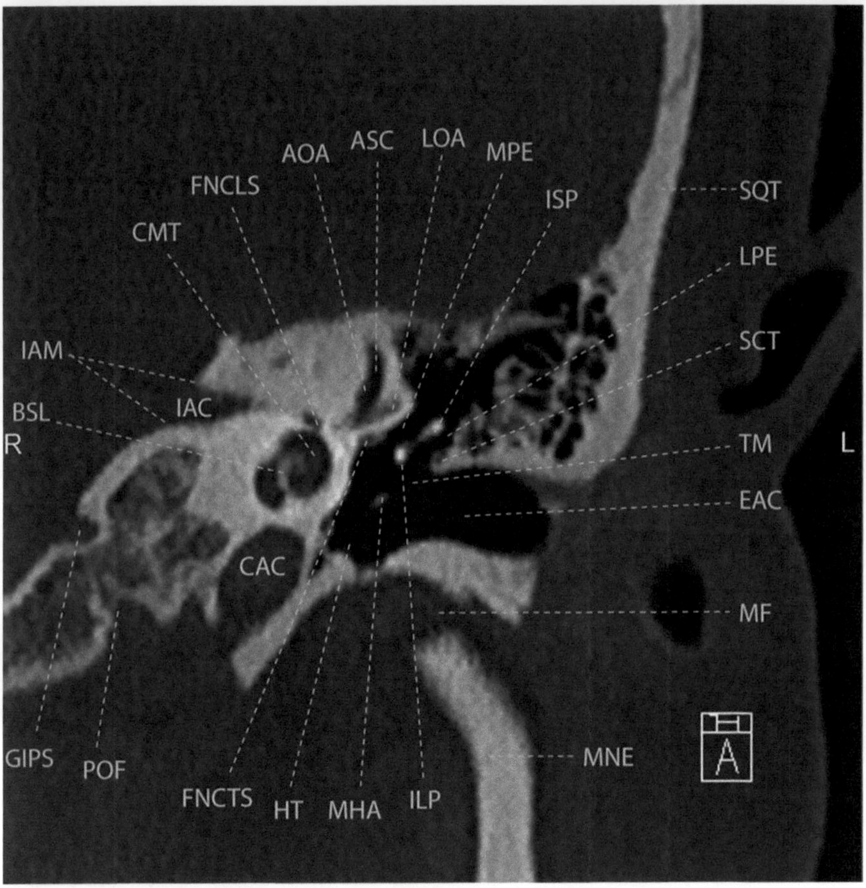

C17: HRCT coronal reformat

AOA	anterior osseous ampulla	IAM	internal auditory canal – meatus
ASC	anterior semicircular canal	ILP	incus – long process
BSL	bony spiral lamina	ISP	incus – short process
CAC	carotid canal	LOA	lateral osseous ampulla
CMT	cochlea – middle turn	LPE	lateral posterior epitympanum
EAC	external auditory canal	MF	mandibular fossa
FNCLS	facial nerve canal – labyrinthine segment	MHA	malleus – handle
FNCTS	facial nerve canal – tympanic segment	MNE	mandible – neck
GIPS	groove of the inferior petrosal sinus	MPE	medial posterior epitympanum
HT	hypotympanum	POF	petrooccipital fissure
IAC	internal auditory canal	SCT	scutum
		SQT	squama temporalis
		TM	tympanic membrane

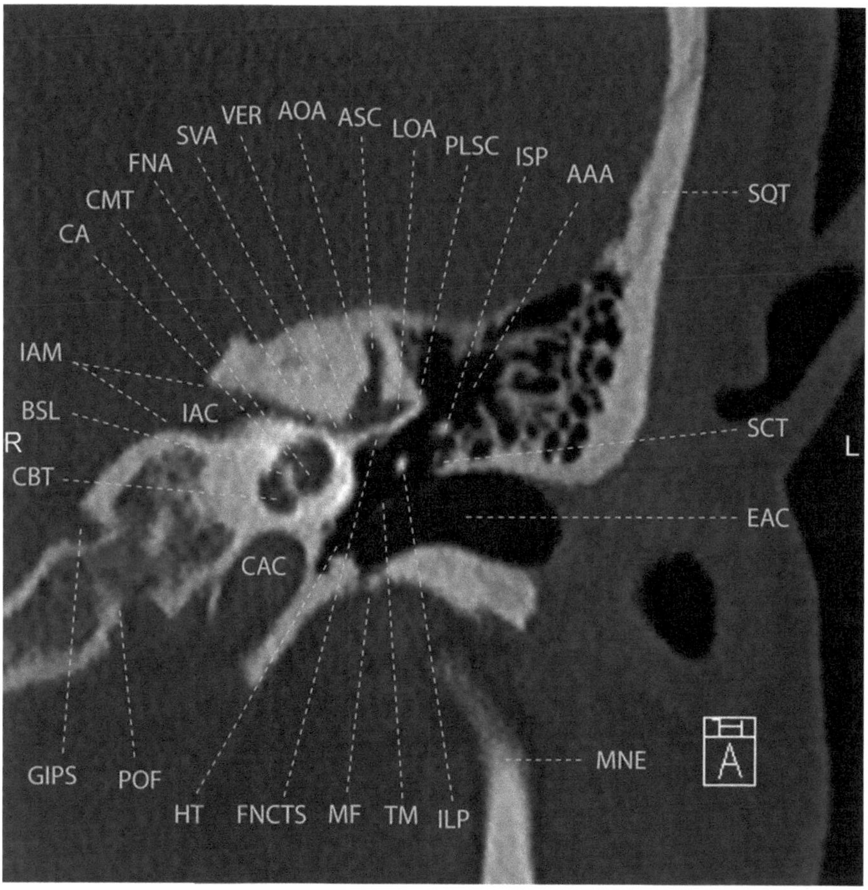

C18: HRCT coronal reformat

AAA	aditus ad antrum	IAC	internal auditory canal
AOA	anterior osseous ampulla	IAM	internal auditory canal – meatus
ASC	anterior semicircular canal	ILP	incus – long process
BSL	bony spiral lamina	ISP	incus – short process
CA	cochlear area	LOA	lateral osseous ampulla
CAC	carotid canal	MF	mandibular fossa
CBT	cochlea – basilar turn	MNE	mandible – neck
CMT	cochlea – middle turn	PLSC	prominence of lateral semicircular canal
EAC	external auditory canal	POF	petrooccipital fissure
FNA	facial nerve area	SCT	scutum
FNCTS	facial nerve canal – tympanic segment	SQT	squama temporalis
GIPS	groove of the inferior petrosal sinus	SVA	superior vestibular area
HT	hypotympanum	TM	tympanic membrane
		VER	vestibule – elliptical recess

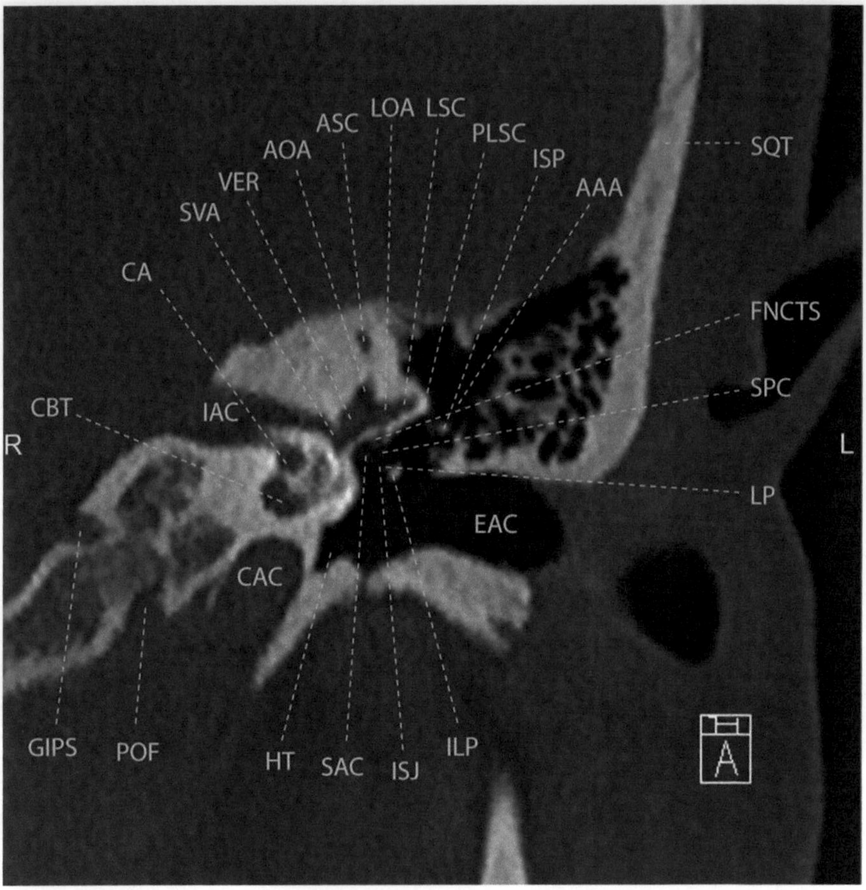

C19: HRCT coronal reformat

AAA	aditus ad antrum
AOA	anterior osseous ampulla
ASC	anterior semicircular canal
CA	cochlear area
CAC	carotid canal
CBT	cochlea – basilar turn
EAC	external auditory canal
FNCTS	facial nerve canal – tympanic segment
GIPS	groove of the inferior petrosal sinus
HT	hypotympanum
IAC	internal auditory canal
ILP	incus – long process
ISJ	incudostapedial joint
ISP	incus – short process
LOA	lateral osseous ampulla
LP	lenticular process
LSC	lateral semicircular canal
PLSC	prominence of lateral semicircular canal
POF	petrooccipital fissure
SAC	stapes – anterior crus
SPC	stapes – posterior crus
SQT	squama temporalis
SVA	superior vestibular area
VER	vestibule – elliptical recess

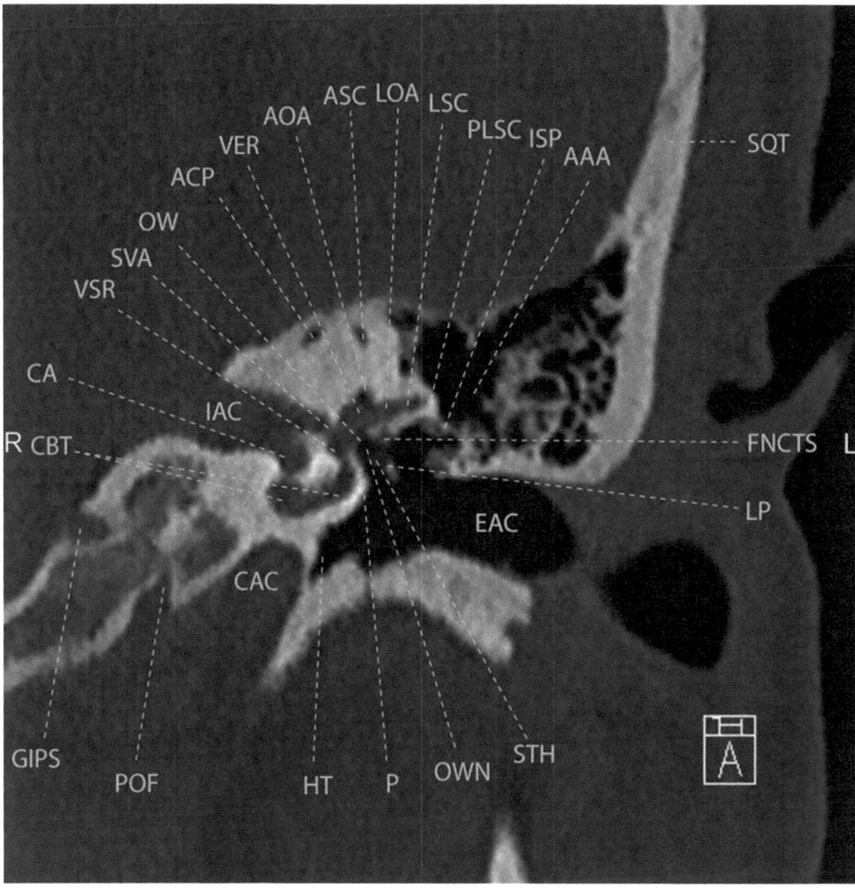

C20: HRCT coronal reformat

AAA	aditus ad antrum
ACP	air cell of the pyramid
AOA	anterior osseous ampulla
ASC	anterior semicircular canal
CA	cochlear area
CAC	carotid canal
CBT	cochlea – basilar turn
EAC	external auditory canal
FNCTS	facial nerve canal – tympanic segment
GIPS	groove of the inferior petrosal sinus
HT	hypotympanum
IAC	internal auditory canal
ISP	incus – short process
LOA	lateral osseous ampulla
LP	lenticular process
LSC	lateral semicircular canal
OW	oval window
OWN	oval window niche
P	promontory
PLSC	prominence of lateral semicircular canal
POF	petrooccipital fissure
SQT	squama temporalis
STH	stapes – head
SVA	superior vestibular area
VER	vestibule – elliptical recess
VSR	vestibule – spherical recess

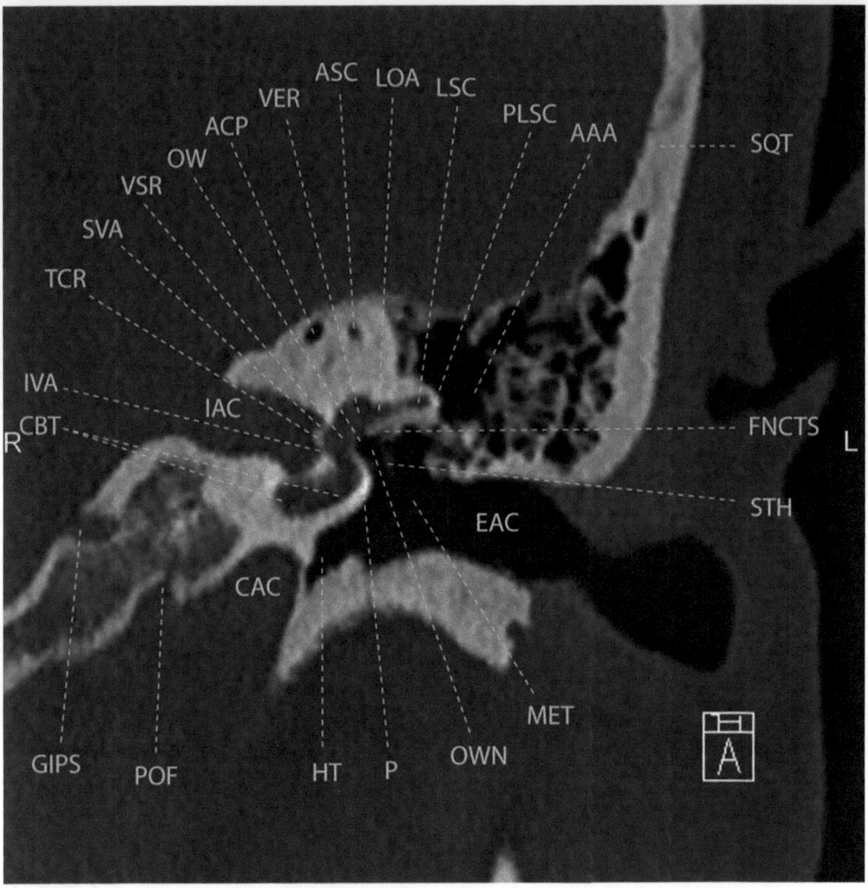

C21: HRCT coronal reformat

AAA	aditus ad antrum	LSC	lateral semicircular canal
ACP	air cell of the pyramid	MET	mesotympanum
ASC	anterior semicircular canal	OW	oval window
CAC	carotid canal	OWN	oval window niche
CBT	cochlea – basilar turn	P	promontory
EAC	external auditory canal	PLSC	prominence of lateral semicircular canal
FNCTS	facial nerve canal – tympanic segment	POF	petrooccipital fissure
GIPS	groove of the inferior petrosal sinus	SQT	squama temporalis
		STH	stapes – head
HT	hypotympanum	SVA	superior vestibular area
IAC	internal auditory canal	TCR	transverse crest
IVA	inferior vestibular area	VER	vestibule – elliptical recess
LOA	lateral osseous ampulla	VSR	vestibule – spherical recess

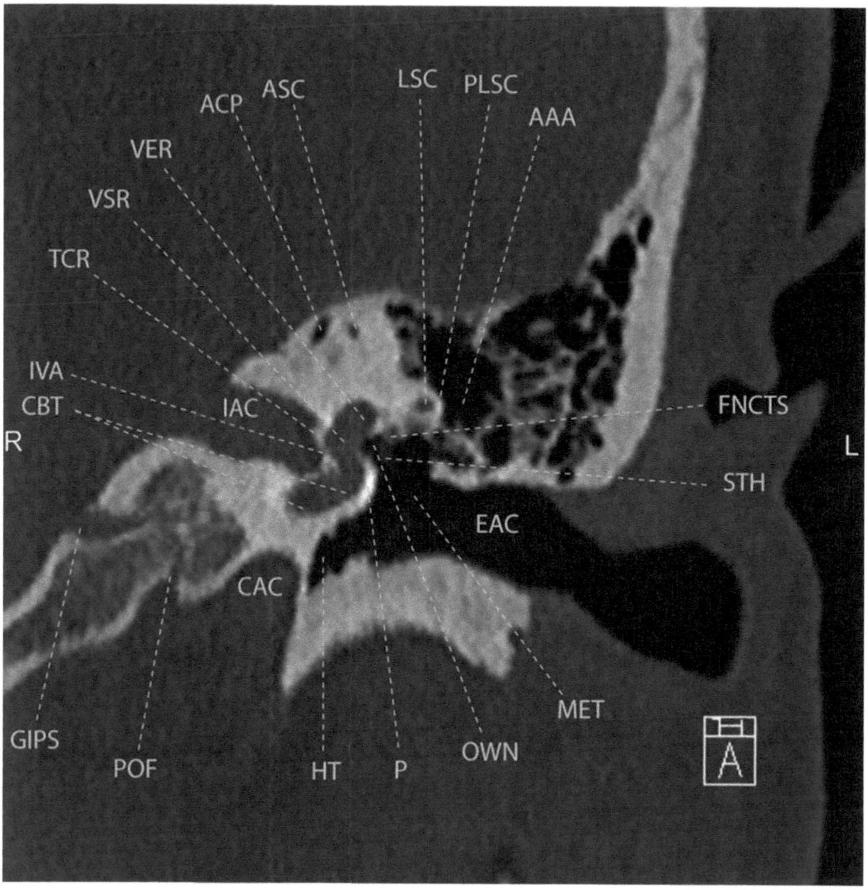

C22: HRCT coronal reformat

AAA	aditus ad antrum
ACP	air cell of the pyramid
ASC	anterior semicircular canal
CAC	carotid canal
CBT	cochlea – basilar turn
EAC	external auditory canal
FNCTS	facial nerve canal – tympanic segment
GIPS	groove of the inferior petrosal sinus
HT	hypotympanum
IAC	internal auditory canal
IVA	inferior vestibular area
LSC	lateral semicircular canal
MET	mesotympanum
OWN	oval window niche
P	promontory
PLSC	prominence of lateral semicircular canal
POF	petrooccipital fissure
STH	stapes head
TCR	transverse crest
VER	vestibule – elliptical recess
VSR	vestibule – spherical recess

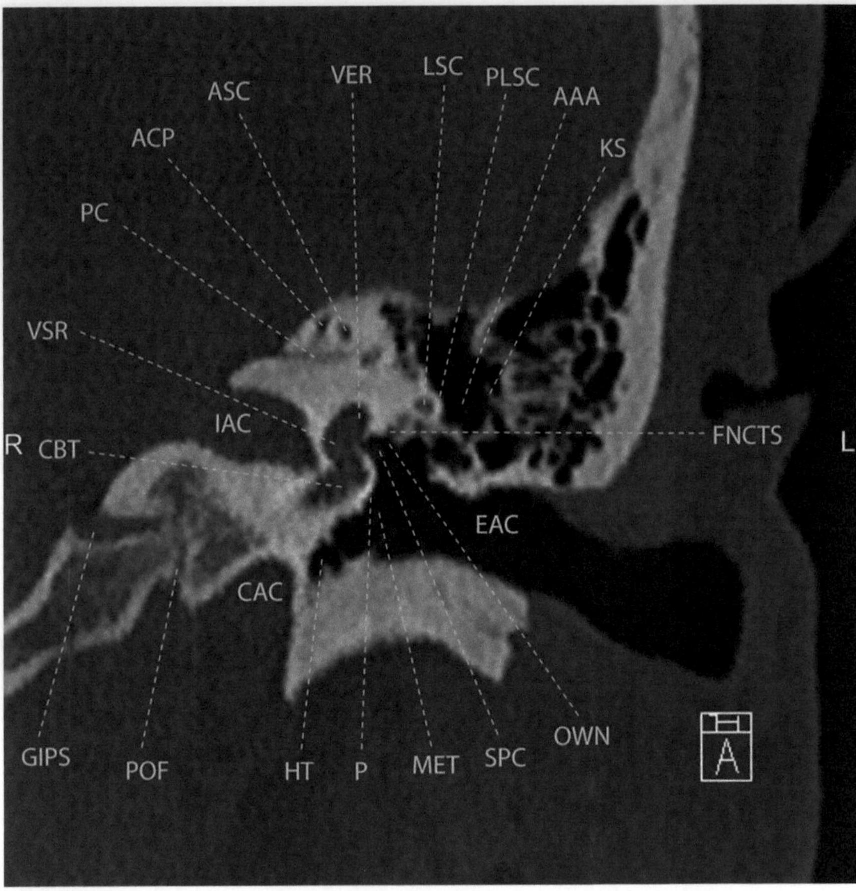

C23: HRCT coronal reformat

AAA	aditus ad antrum	KS	Koerner's septum
ACP	air cell of the pyramid	LSC	lateral semicircular canal
ASC	anterior semicircular canal	MET	mesotympanum
CAC	carotid canal	OWN	oval window niche
CBT	cochlea – basilar turn	P	promontory
EAC	external auditory canal	PC	petromastoid canal
FNCTS	facial nerve canal – tympanic segment	PLSC	prominence of lateral semicircular canal
GIPS	groove of the inferior petrosal sinus	POF	petrooccipital fissure
		SPC	stapes posterior crus
HT	hypotympanum	VER	vestibule – elliptical recess
IAC	internal auditory canal	VSR	vestibule – spherical recess

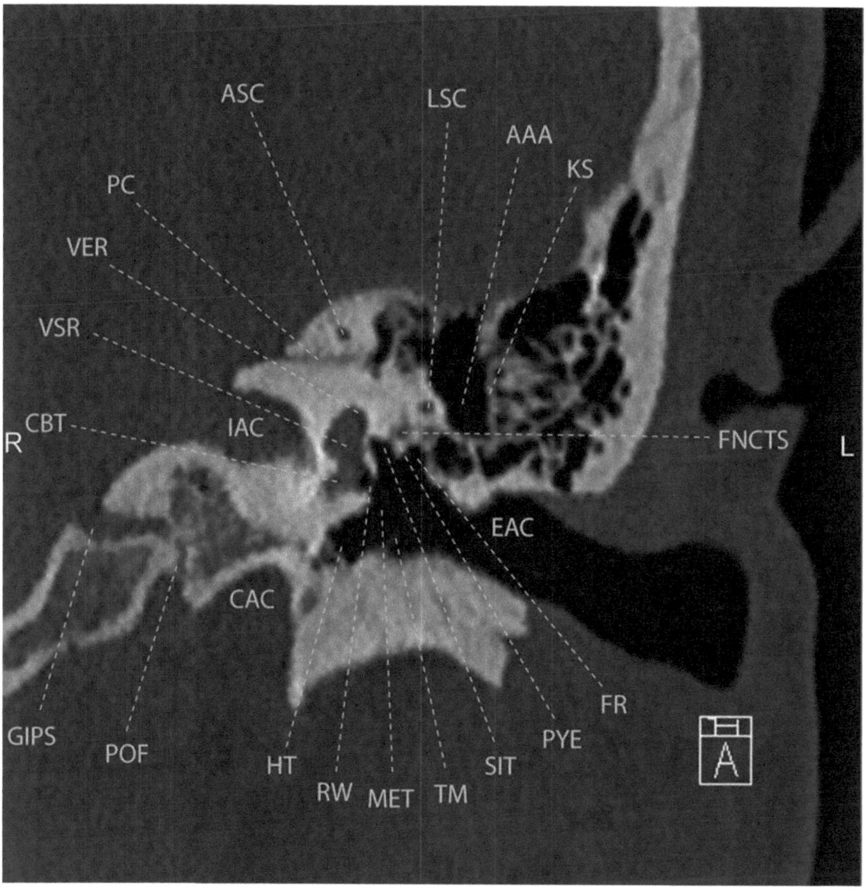

C24: HRCT coronal reformat

AAA	aditus ad antrum	KS	Koerner's septum
ASC	anterior semicircular canal	LSC	lateral semicircular canal
CAC	carotid canal	MET	mesotympanum
CBT	cochlea – basilar turn	PC	petromastoid canal
EAC	external auditory canal	POF	petrooccipital fissure
FNCTS	facial nerve canal – tympanic segment	PYE	pyramidal eminence
		RW	round window
FR	facial recess	SIT	sinus tympani
GIPS	groove of the inferior petrosal sinus	TM	tympanic membrane
		VER	vestibule – elliptical recess
HT	hypotympanum	VSR	vestibule – spherical recess
IAC	internal auditory canal		

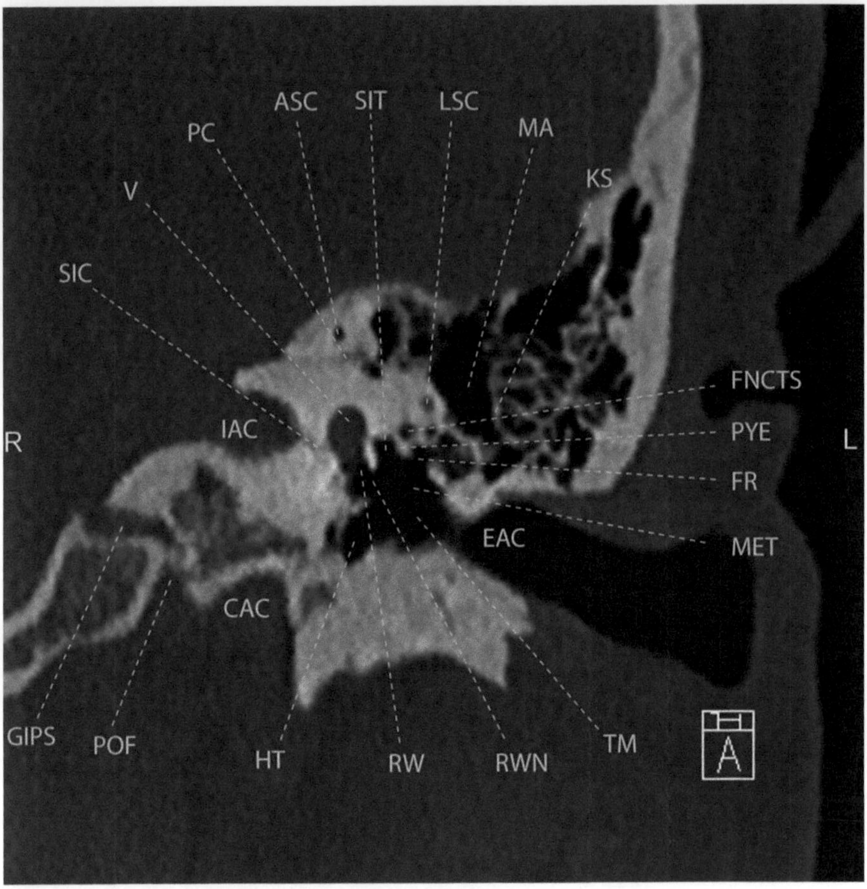

C25: HRCT coronal reformat

ASC	anterior semicircular canal	MA	mastoid antrum
CAC	carotid canal	MET	mesotympanum
EAC	external auditory canal	PC	petromastoid canal
FNCTS	facial nerve canal – tympanic segment	POF	petrooccipital fissure
FR	facial recess	PYE	pyramidal eminence
GIPS	groove of the inferior petrosal sinus	RW	round window
HT	hypotympanum	RWN	round window niche
IAC	internal auditory canal	SIC	singular canal
KS	Koerner's septum	SIT	sinus tympani
LSC	lateral semicircular canal	TM	tympanic membrane
		V	vestibule

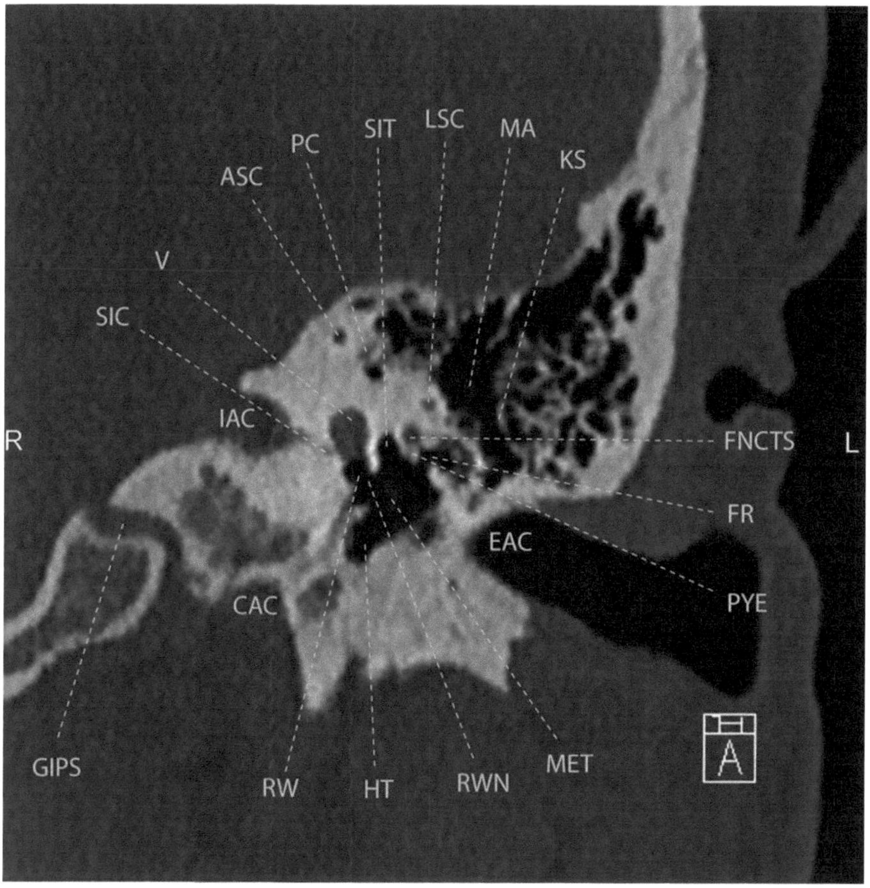

C26: HRCT coronal reformat

ASC	anterior semicircular canal	LSC	lateral semicircular canal
CAC	carotid canal	MA	mastoid antrum
EAC	external auditory canal	MET	mesotympanum
FNCTS	facial nerve canal – tympanic segment	PC	petromastoid canal
FR	facial recess	PYE	pyramidal eminence
GIPS	groove of the inferior petrosal sinus	RW	round window
HT	hypotympanum	RWN	round window niche
IAC	internal auditory canal	SIC	singular canal
KS	Koerner's septum	SIT	sinus tympani
		V	vestibule

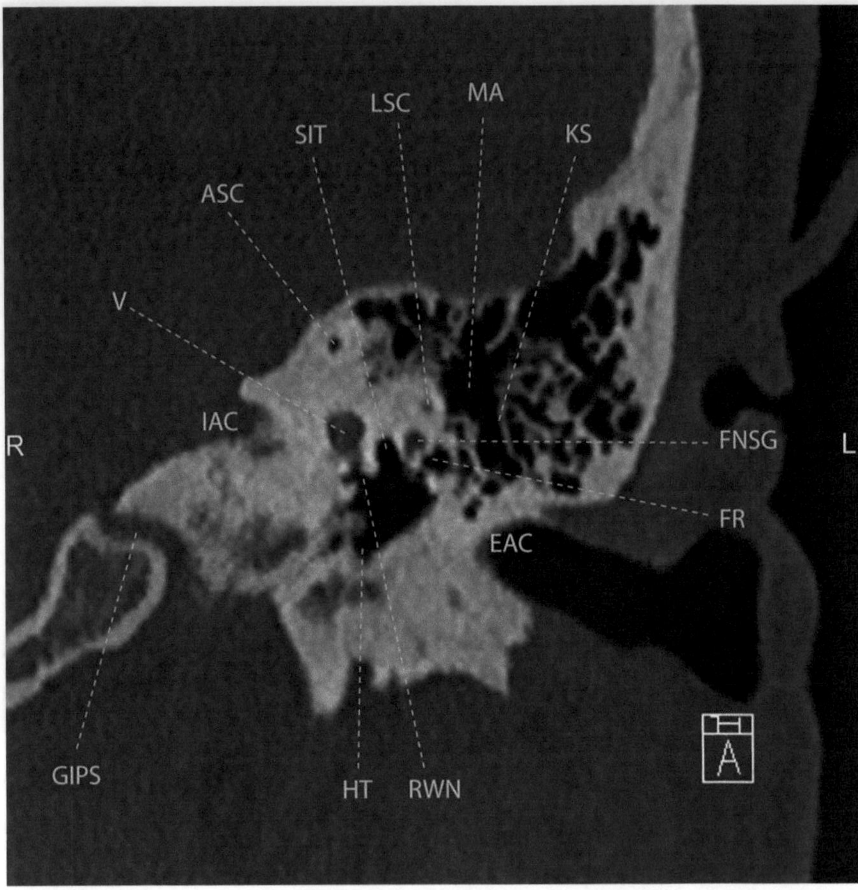

C27: HRCT coronal reformat

ASC	anterior semicircular canal	IAC	internal auditory canal
EAC	external auditory canal	KS	Koerner's septum
FNSG	facial nerve canal – second genu	LSC	lateral semicircular canal
		MA	mastoid antrum
FR	facial recess	RWN	round window niche
GIPS	groove of the inferior petrosal sinus	SIT	sinus tympani
		V	vestibule
HT	hypotympanum		

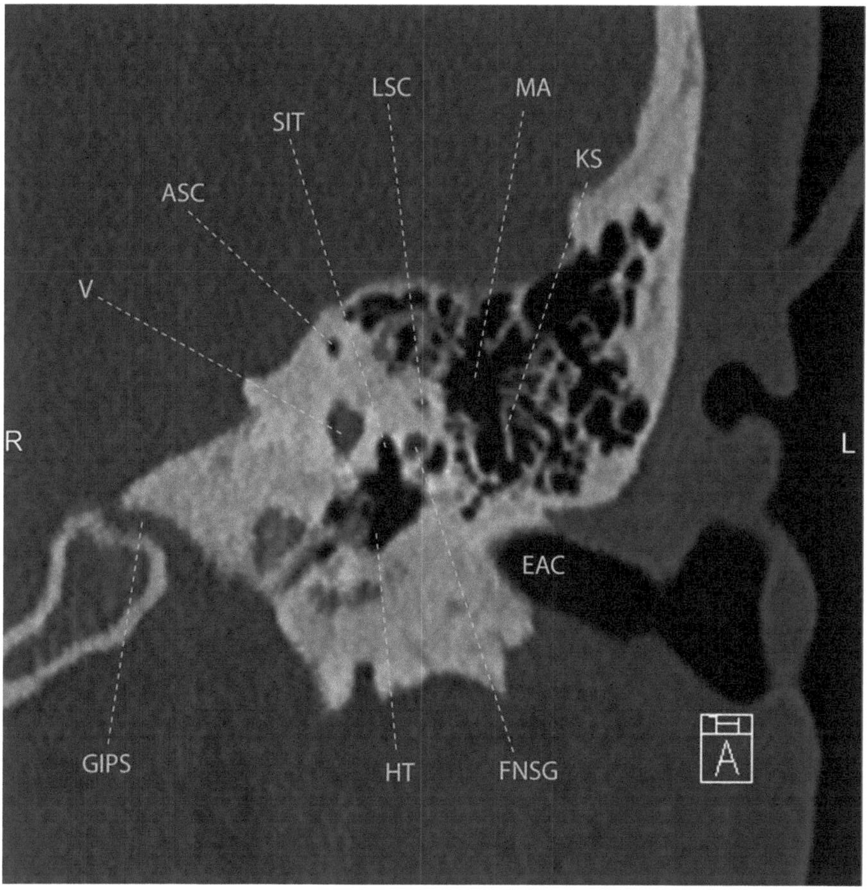

C28: HRCT coronal reformat

ASC	anterior semicircular canal	HT	hypotympanum
EAC	external auditory canal	KS	Koerner's septum
FNSG	facial nerve canal – second genu	LSC	lateral semicircular canal
		MA	mastoid antrum
GIPS	groove of the inferior petrosal sinus	SIT	sinus tympani
		V	vestibule

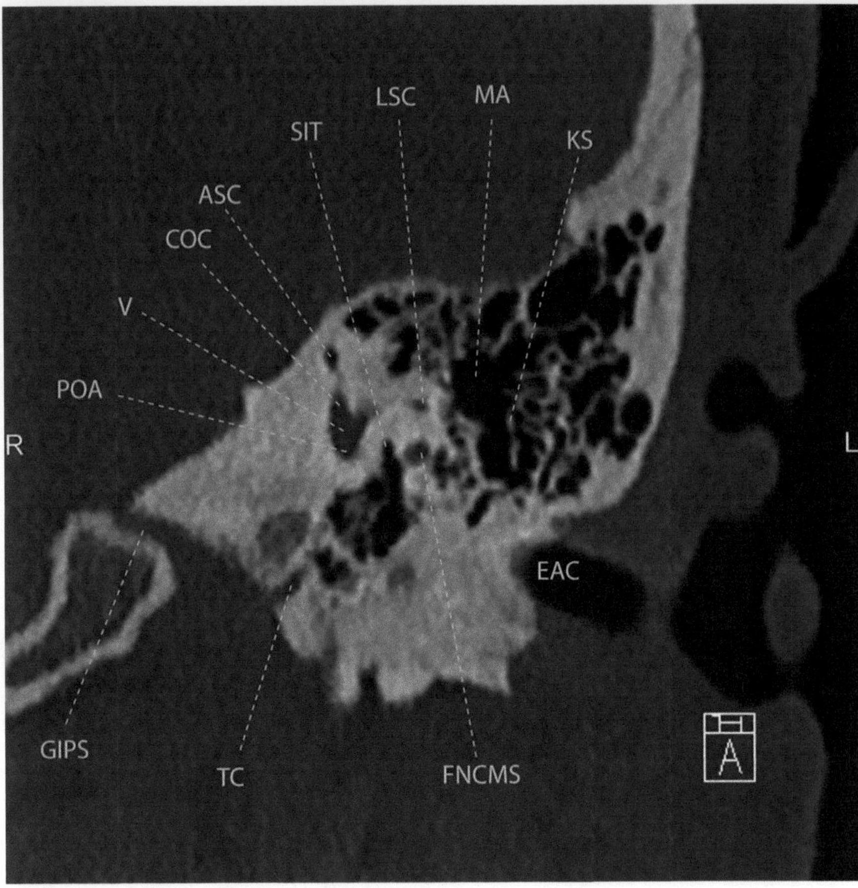

C29: HRCT coronal reformat

ASC	anterior semicircular canal	KS	Koerner's septum
COC	common osseous crus	LSC	lateral semicircular canal
EAC	external auditory canal	MA	mastoid antrum
FNCMS	facial nerve canal – mastoid segment	POA	posterior osseous ampulla
GIPS	groove of the inferior petrosal sinus	SIT	sinus tympani
		TC	tympanic canaliculus
		V	vestibule

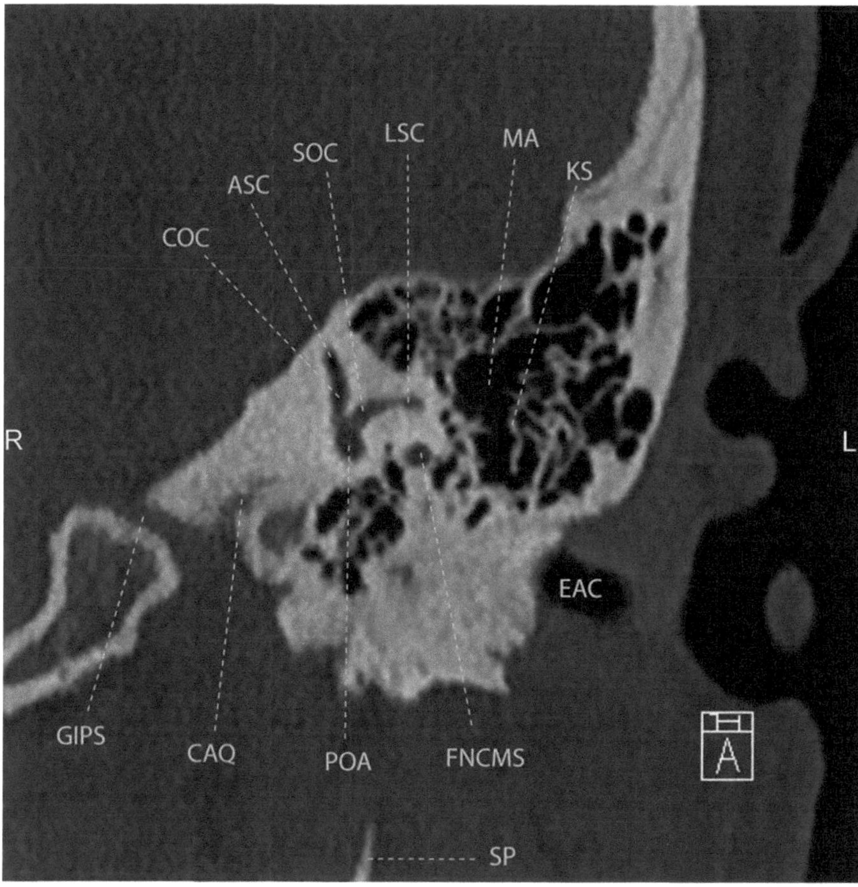

C30: HRCT coronal reformat

ASC	anterior semicircular canal	KS	Koerner's septum
CAQ	cochlear aqueduct	LSC	lateral semicircular canal
COC	common osseous crus	MA	mastoid antrum
EAC	external auditory canal	POA	posterior osseous ampulla
FNCMS	facial nerve canal – mastoid segment	SOC	simple osseous crus
GIPS	groove of the inferior petrosal sinus	SP	styloid process

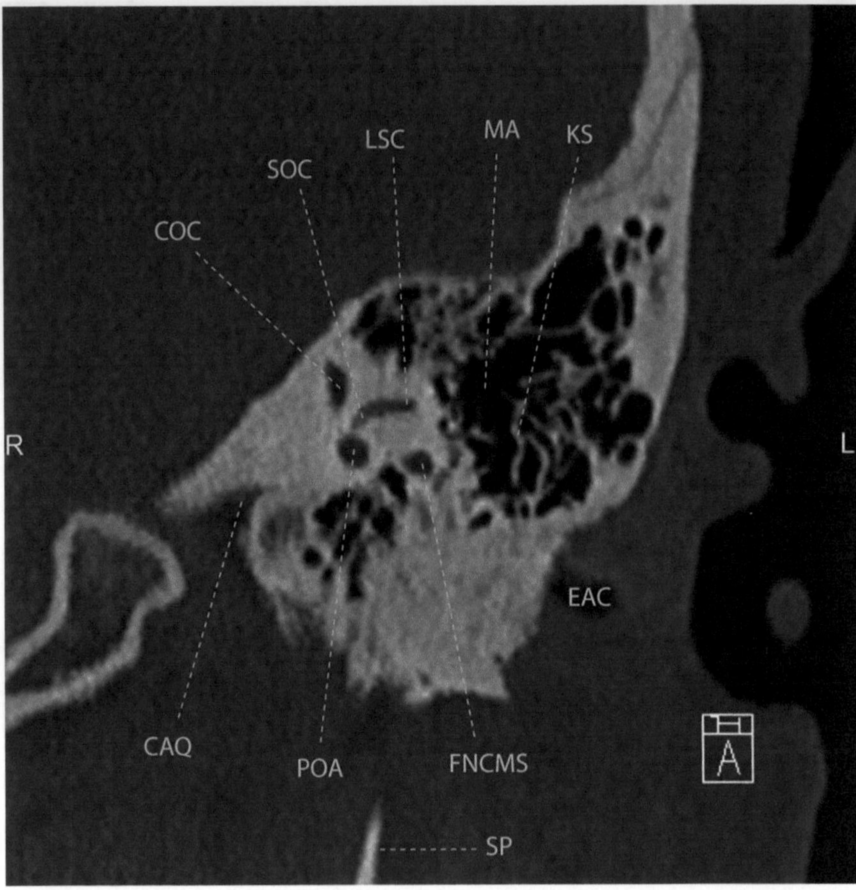

C31: HRCT coronal reformat

CAQ	cochlear aqueduct	LSC	lateral semicircular canal
COC	common osseous crus	MA	mastoid antrum
EAC	external auditory canal	POA	posterior osseous ampulla
FNCMS	facial nerve canal – mastoid segment	SOC	simple osseous crus
		SP	styloid process
KS	Koerner's septum		

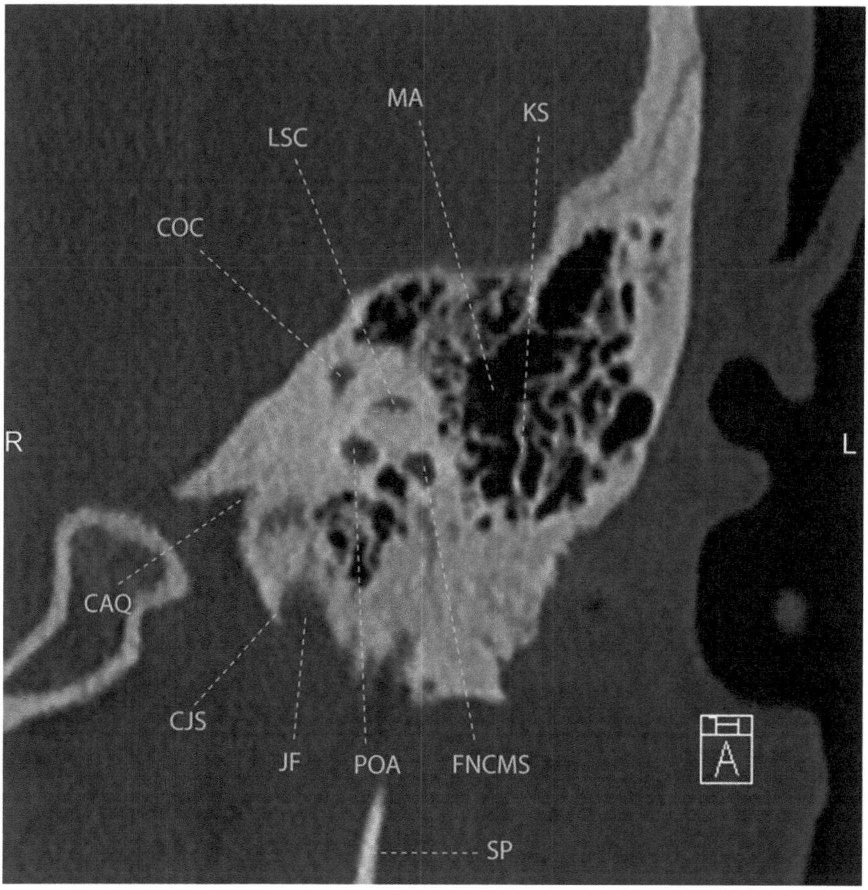

C32: HRCT coronal reformat

CAQ	cochlear aqueduct	KS	Koerner's septum
CJS	caroticojugular spine	LSC	lateral semicircular canal
COC	common osseous crus	MA	mastoid antrum
FNCMS	facial nerve canal – mastoid segment	POA	posterior osseous ampulla
JF	jugular foramen	SP	styloid process

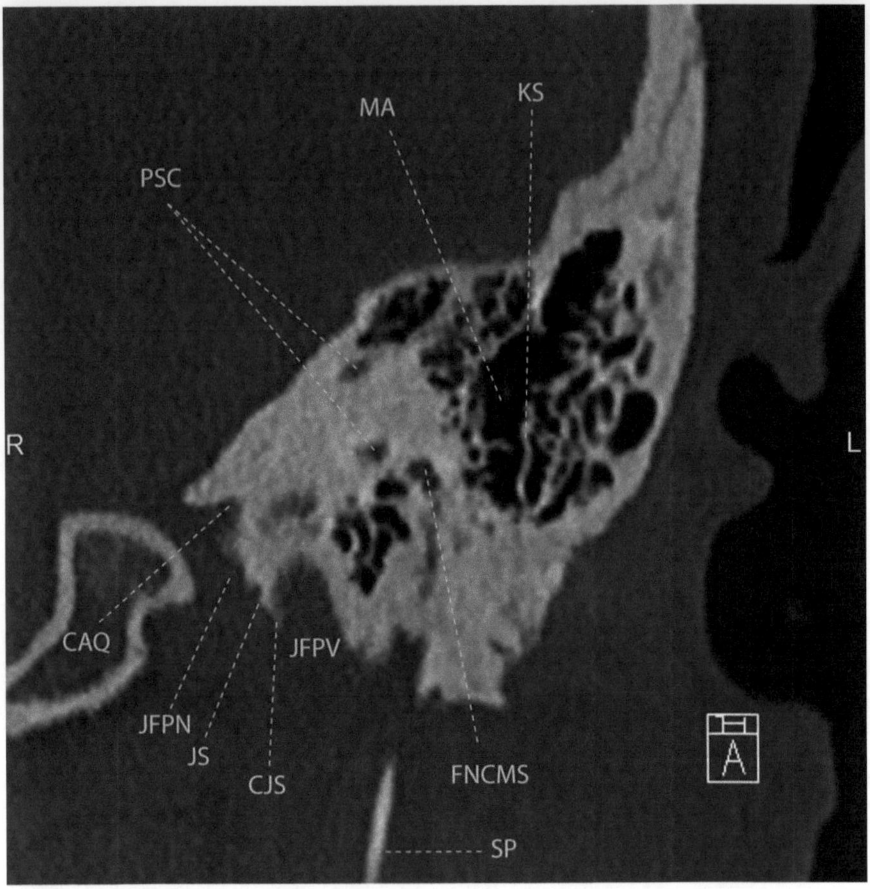

C33: HRCT coronal reformat

CAQ	cochlear aqueduct
CJS	caroticojugular spine
FNCMS	facial nerve canal – mastoid segment
JFPN	jugular foramen – pars nervosa
JFPV	jugular foramen – pars vascularis
JS	jugular spine
KS	Koerner's septum
MA	mastoid antrum
PSC	posterior semicircular canal
SP	styloid process

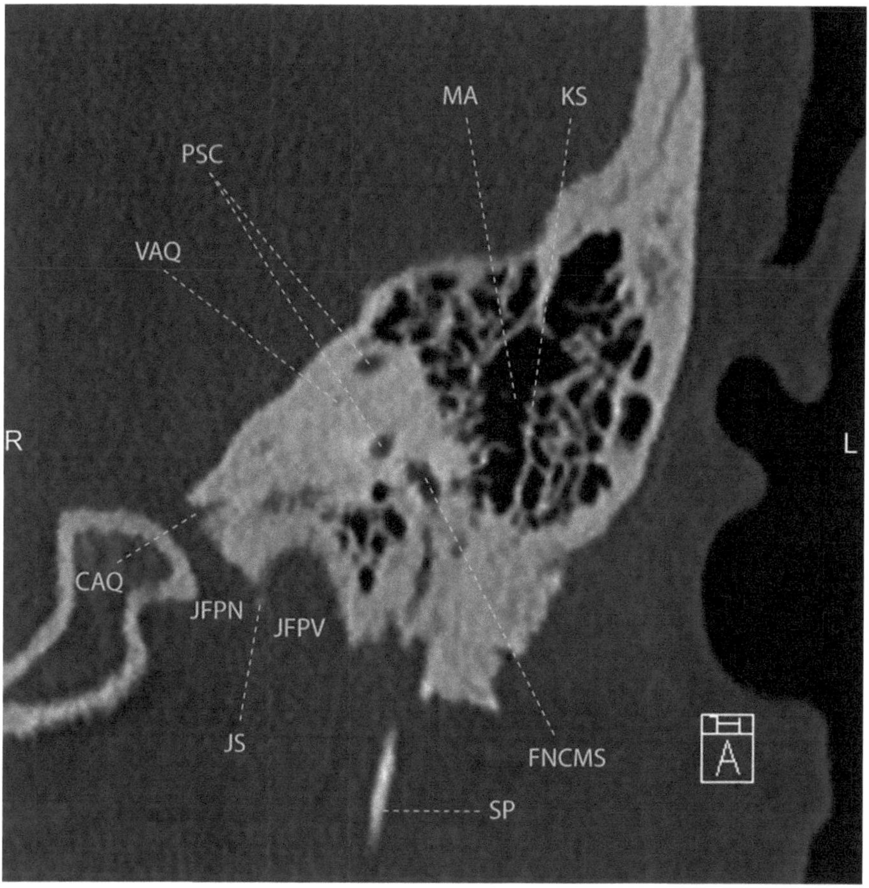

C34: HRCT coronal reformat

CAQ	cochlear aqueduct	JS	jugular spine
FNCMS	facial nerve canal – mastoid segment	KS	Koerner's septum
		MA	mastoid antrum
JFPN	jugular foramen – pars nervosa	PSC	posterior semicircular canal
		SP	styloid process
JFPV	jugular foramen – pars vascularis	VAQ	vestibular aqueduct

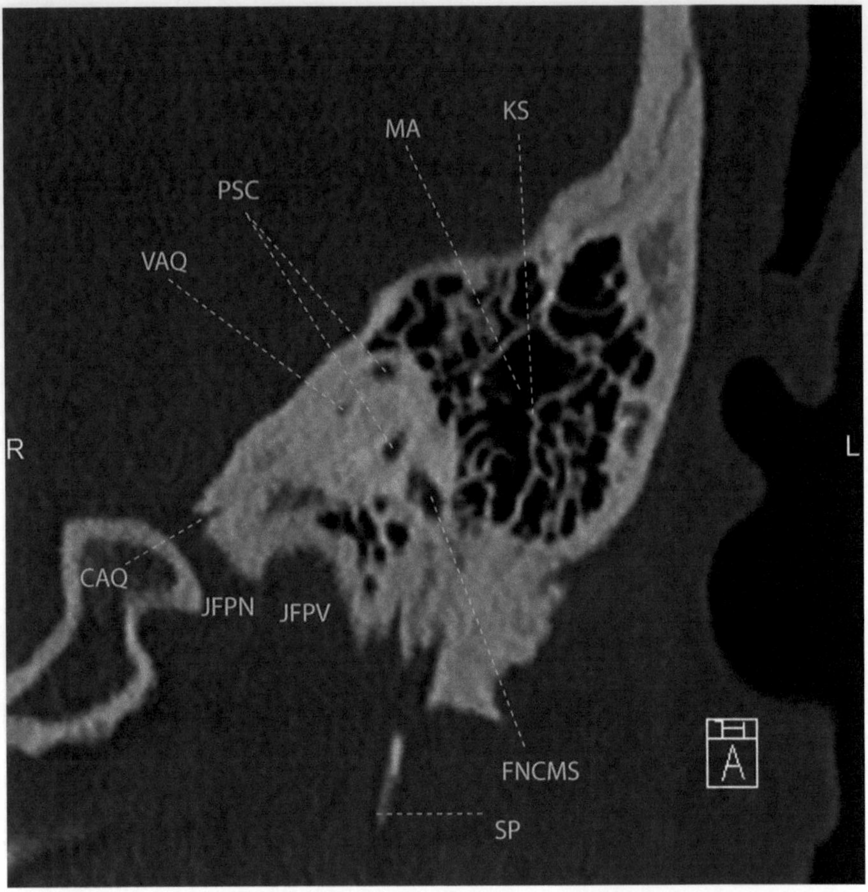

C35: HRCT coronal reformat

CAQ	cochlear aqueduct	KS	Koerner's septum
FNCMS	facial nerve canal – mastoid segment	MA	mastoid antrum
JFPN	jugular foramen – pars nervosa	PSC	posterior semicircular canal
JFPV	jugular foramen – pars vascularis	SP	styloid process
		VAQ	vestibular aqueduct

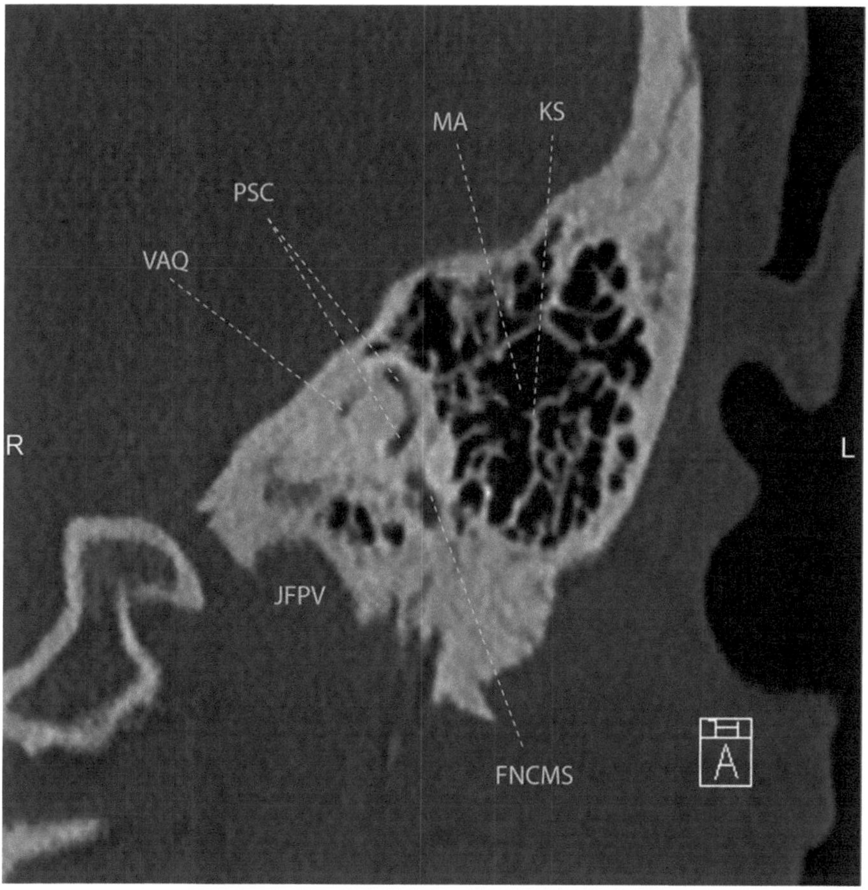

C36: HRCT coronal reformat

FNCMS	facial nerve canal – mastoid segment	KS	Koerner's septum
		MA	mastoid antrum
JFPV	jugular foramen – pars vascularis	PSC	posterior semicircular canal
		VAQ	vestibular aqueduct

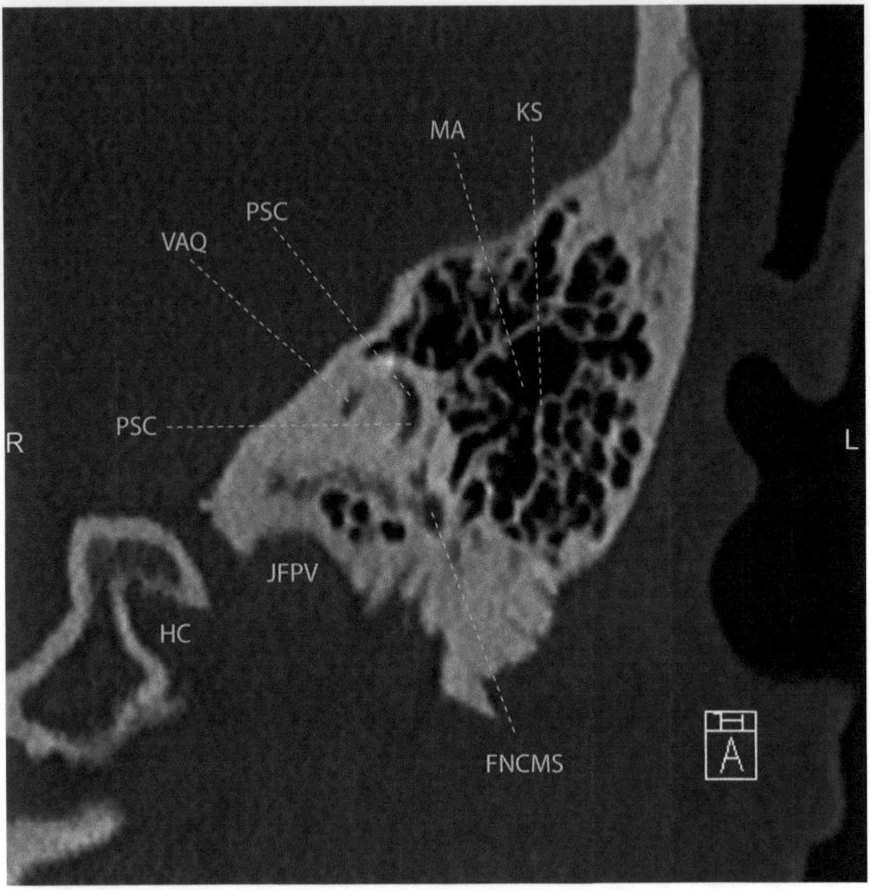

C37: HRCT coronal reformat

FNCMS	facial nerve canal – mastoid segment	KS	Koerner's septum
HC	hypoglossal canal	MA	mastoid antrum
JFPV	jugular foramen – pars vascularis	PSC	posterior semicircular canal
		VAQ	vestibular aqueduct

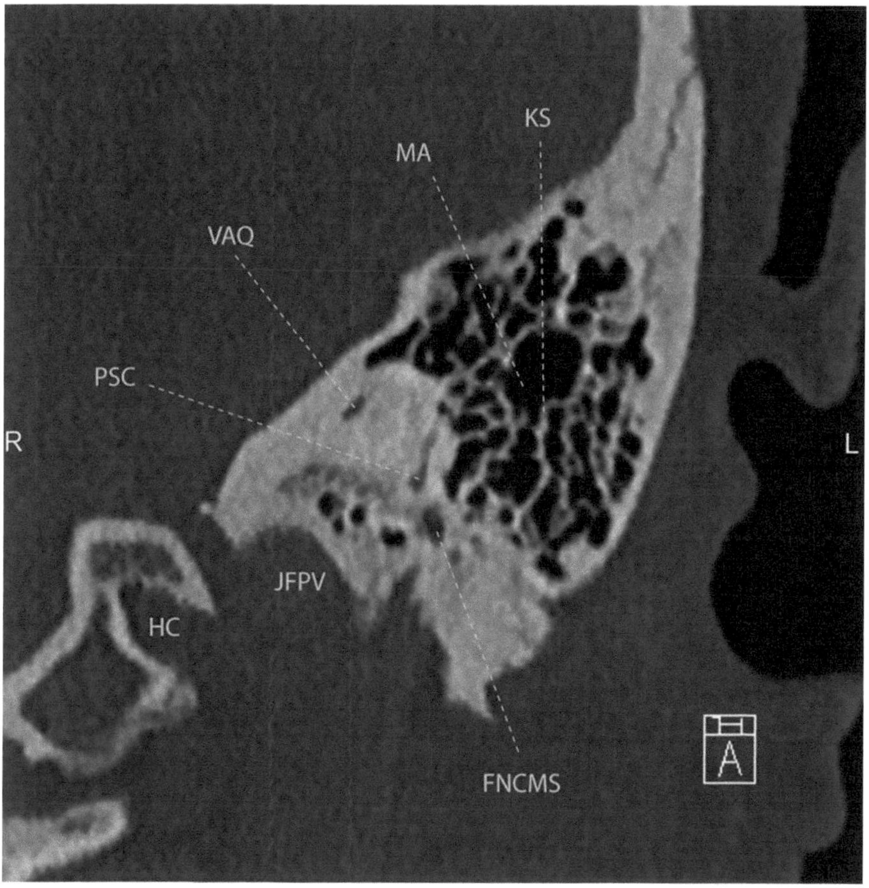

C38: HRCT coronal reformat

FNCMS	facial nerve canal – mastoid segment	KS	Koerner's septum
HC	hypoglossal canal	MA	mastoid antrum
JFPV	jugular foramen – pars vascularis	PSC	posterior semicircular canal
		VAQ	vestibular aqueduct

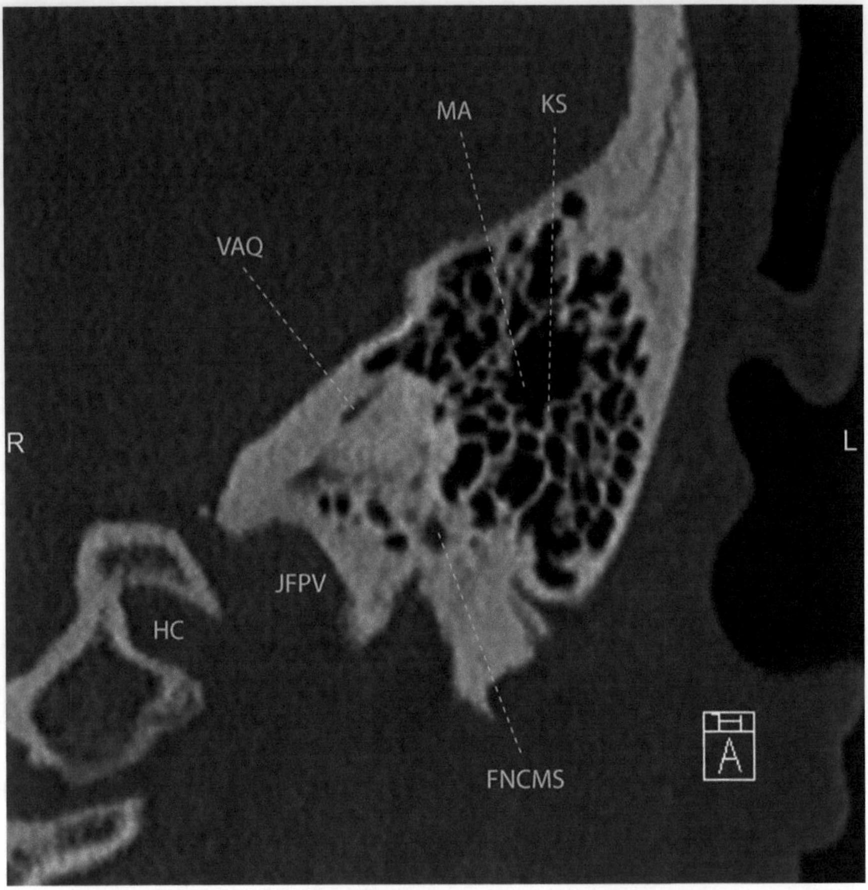

C39: HRCT coronal reformat

FNCMS	facial nerve canal – mastoid segment
HC	hypoglossal canal
JFPV	jugular foramen – pars vascularis
KS	Koerner's septum
MA	mastoid antrum
VAQ	vestibular aqueduct

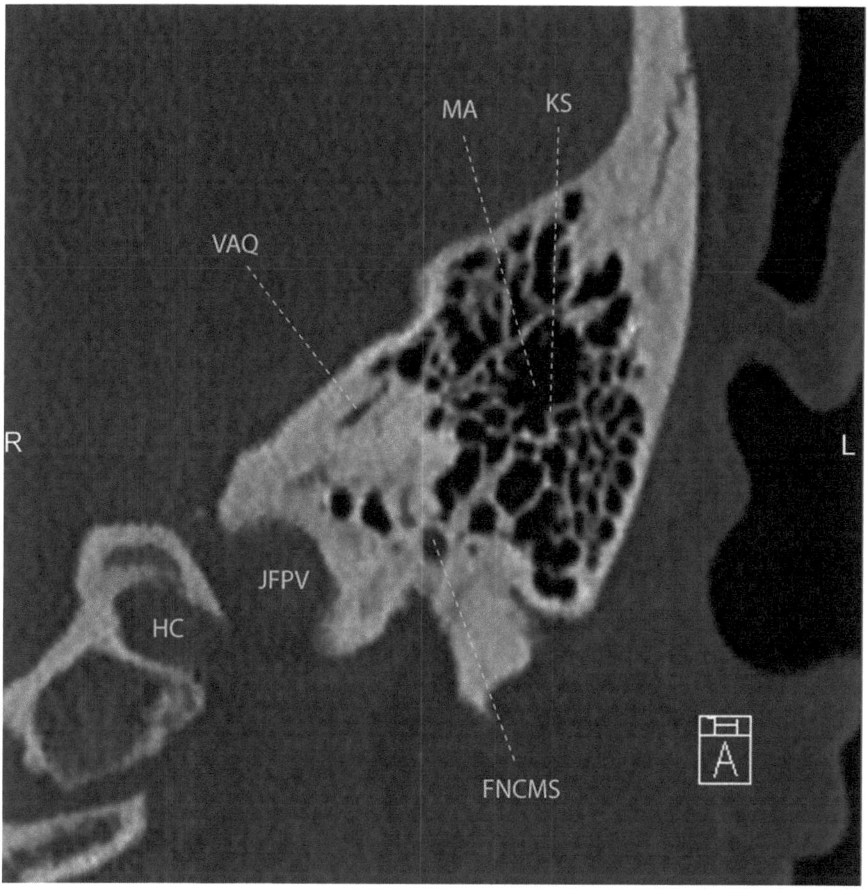

C40: HRCT coronal reformat

FNCMS	facial nerve canal – mastoid segment	KS	Koerner's septum
HC	hypoglossal canal	MA	mastoid antrum
JFPV	jugular foramen – pars vascularis	VAQ	vestibular aqueduct

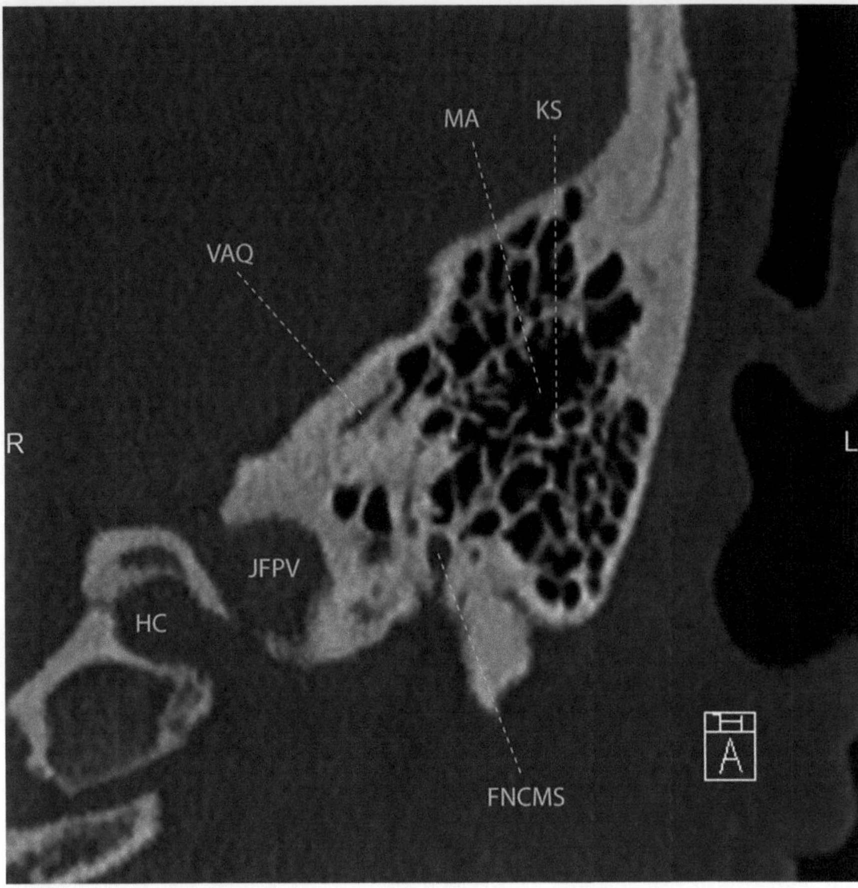

C41: HRCT coronal reformat

FNCMS	facial nerve canal – mastoid segment	KS	Koerner's septum
HC	hypoglossal canal	MA	mastoid antrum
JFPV	jugular foramen – pars vascularis	VAQ	vestibular aqueduct

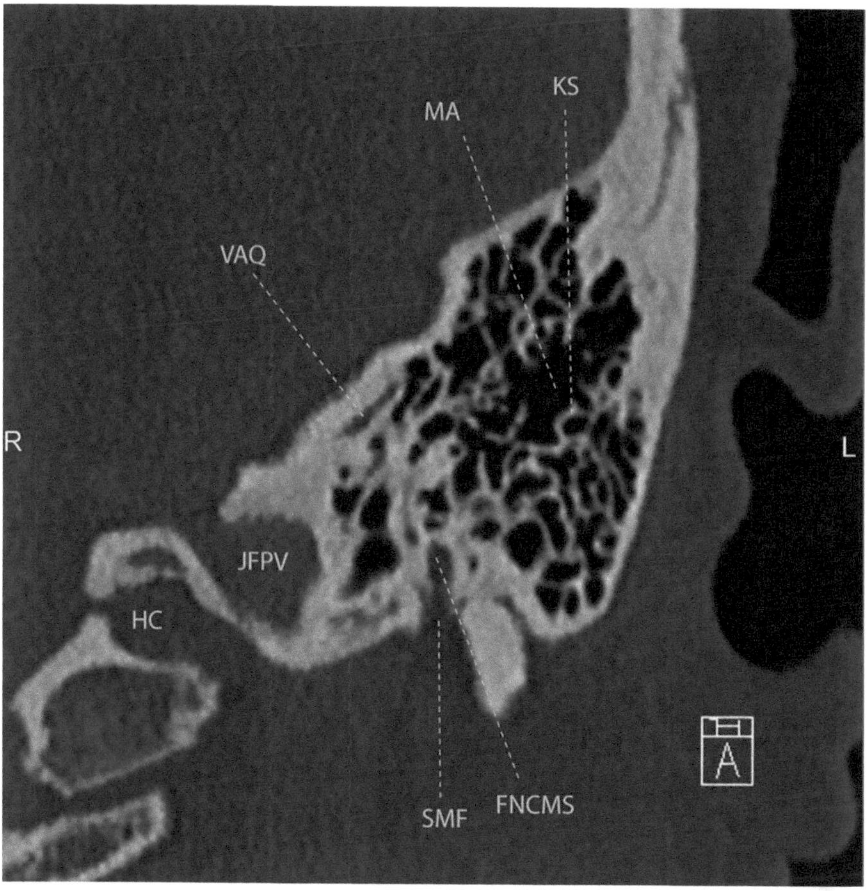

C42: HRCT coronal reformat

FNCMS	facial nerve canal – mastoid segment	KS	Koerner's septum
HC	hypoglossal canal	MA	mastoid antrum
JFPV	jugular foramen – pars vascularis	SMF	stylomastoid foramen
		VAQ	vestibular aqueduct

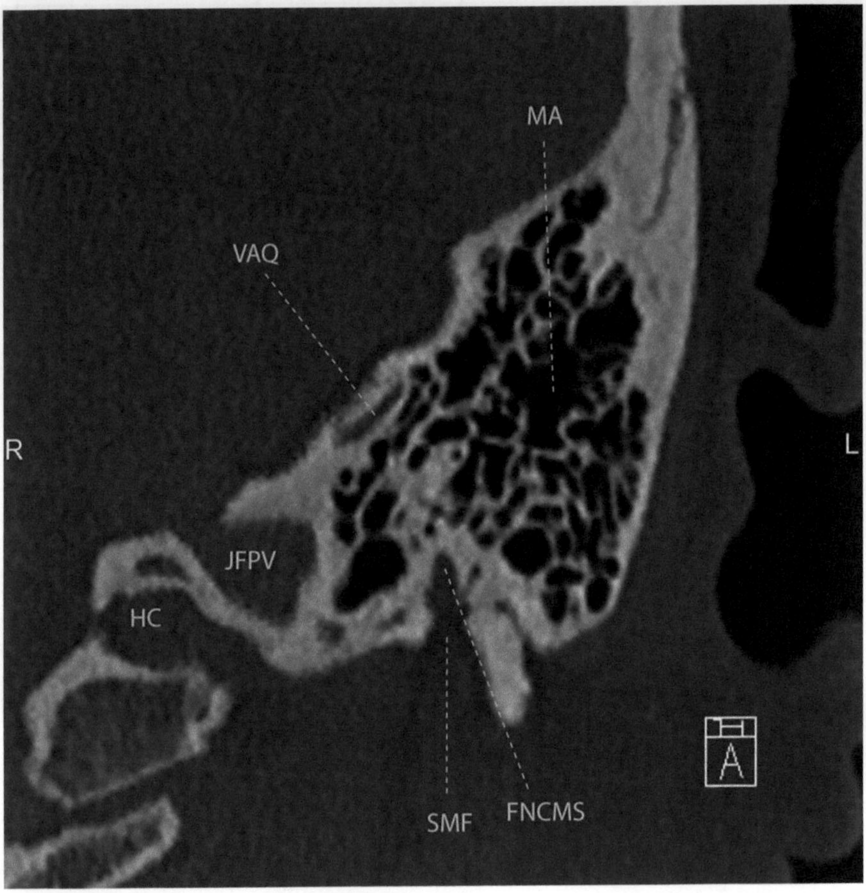

C43: HRCT coronal reformat

FNCMS	facial nerve canal – mastoid segment
HC	hypoglossal canal
JFPV	jugular foramen – pars vascularis
MA	mastoid antrum
SMF	stylomastoid foramen
VAQ	vestibular aqueduct

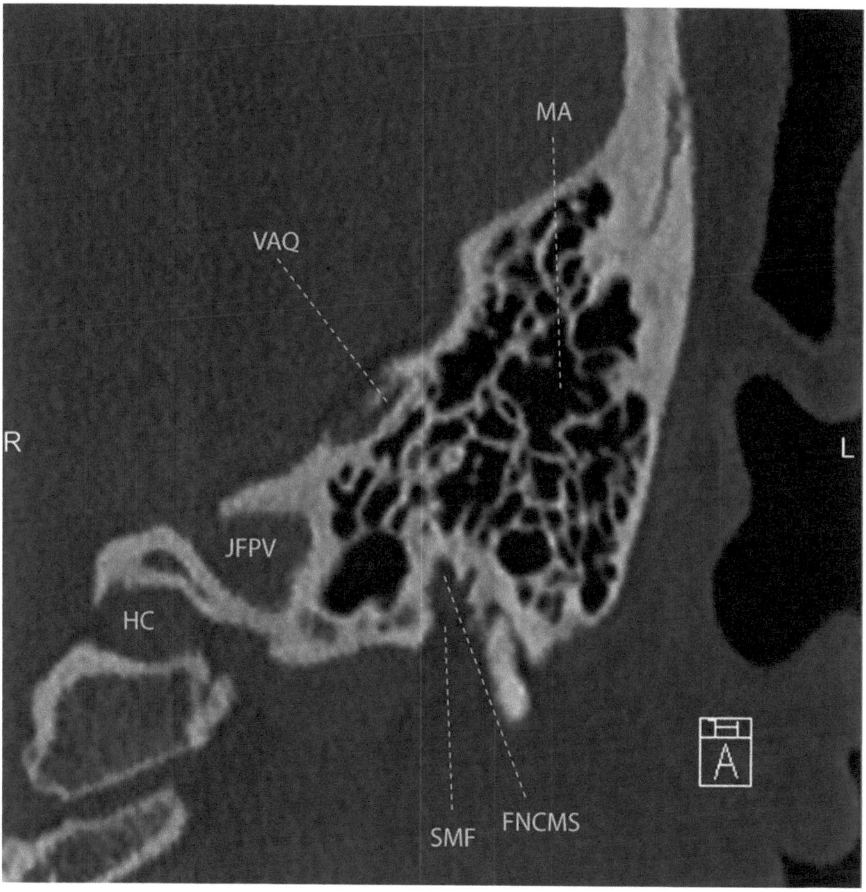

C44: HRCT coronal reformat

FNCMS	facial nerve canal – mastoid segment	MA	mastoid antrum
HC	hypoglossal canal	SMF	stylomastoid foramen
JFPV	jugular foramen – pars vascularis	VAQ	vestibular aqueduct

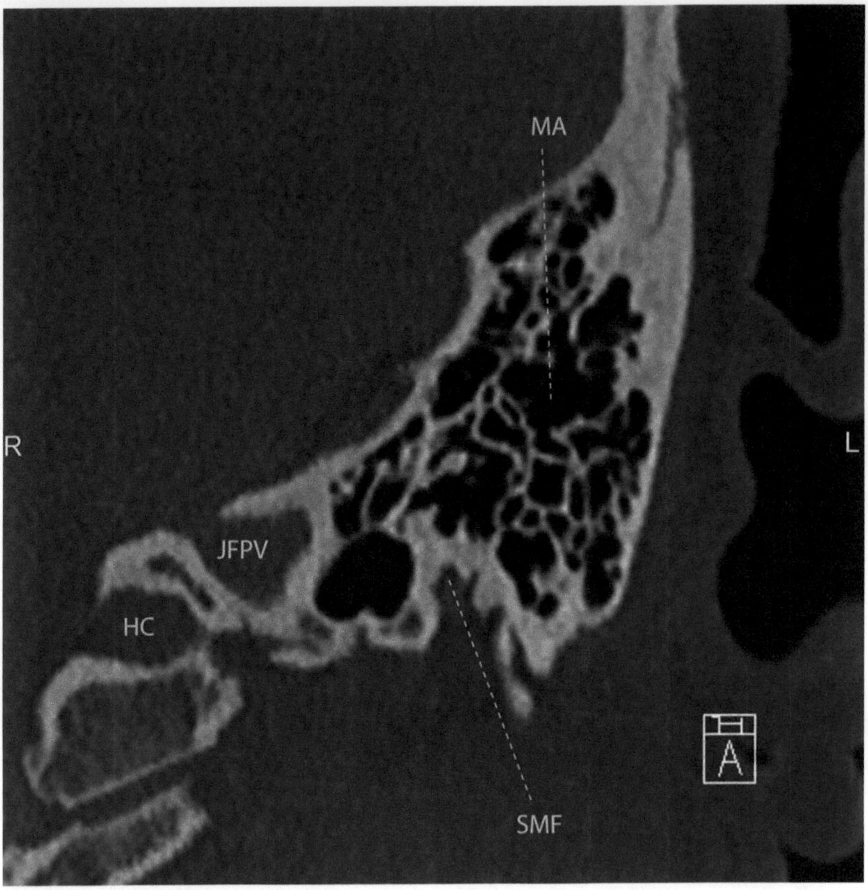

C45: HRCT coronal reformat

HC	hypoglossal canal	MA	mastoid antrum
JFPV	jugular foramen – pars vascularis	SMF	stylomastoid foramen

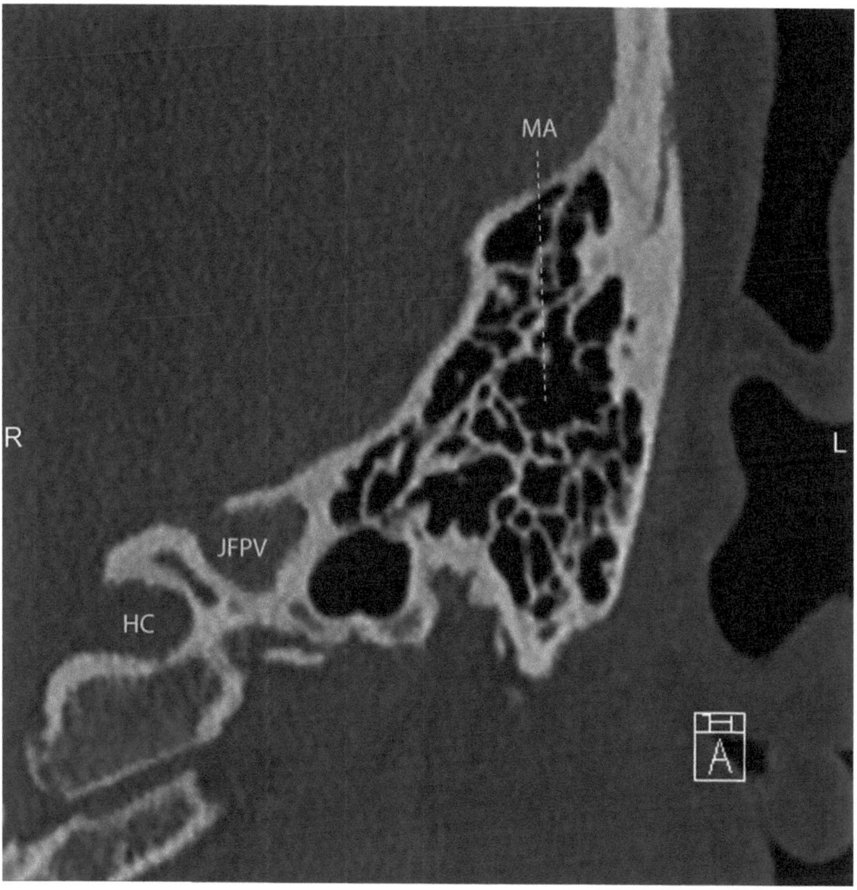

C46: HRCT coronal reformat

HC	hypoglossal canal	MA	mastoid antrum
JFPV	jugular foramen – pars vascularis		

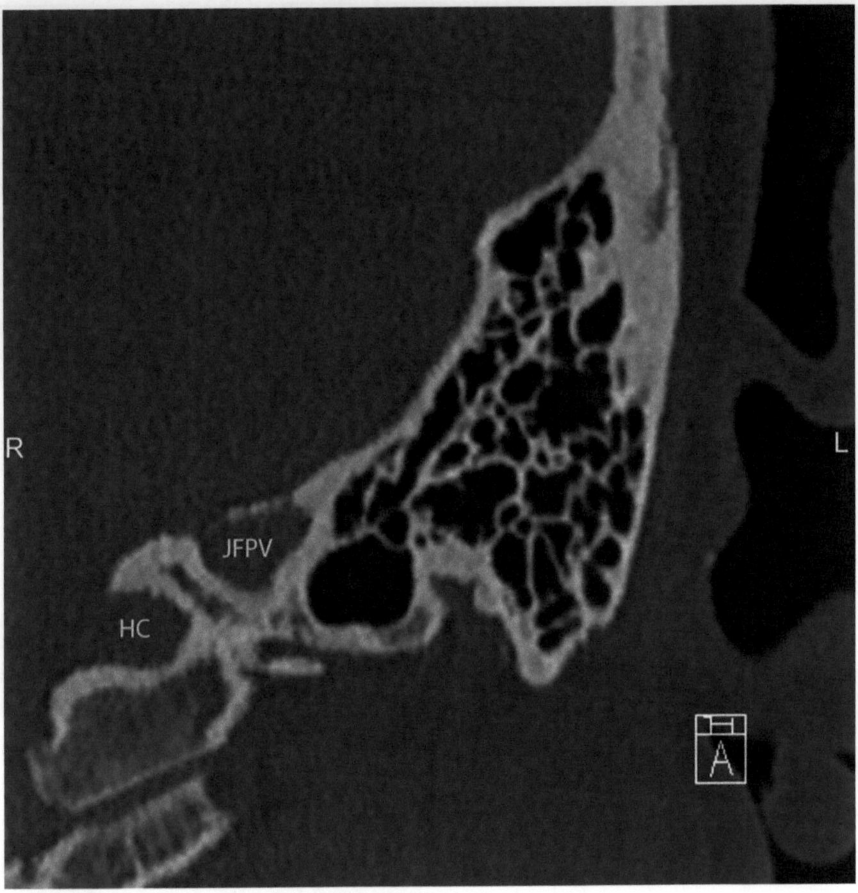

C47: HRCT coronal reformat

HC hypoglossal canal
JFPV jugular foramen – pars vascularis

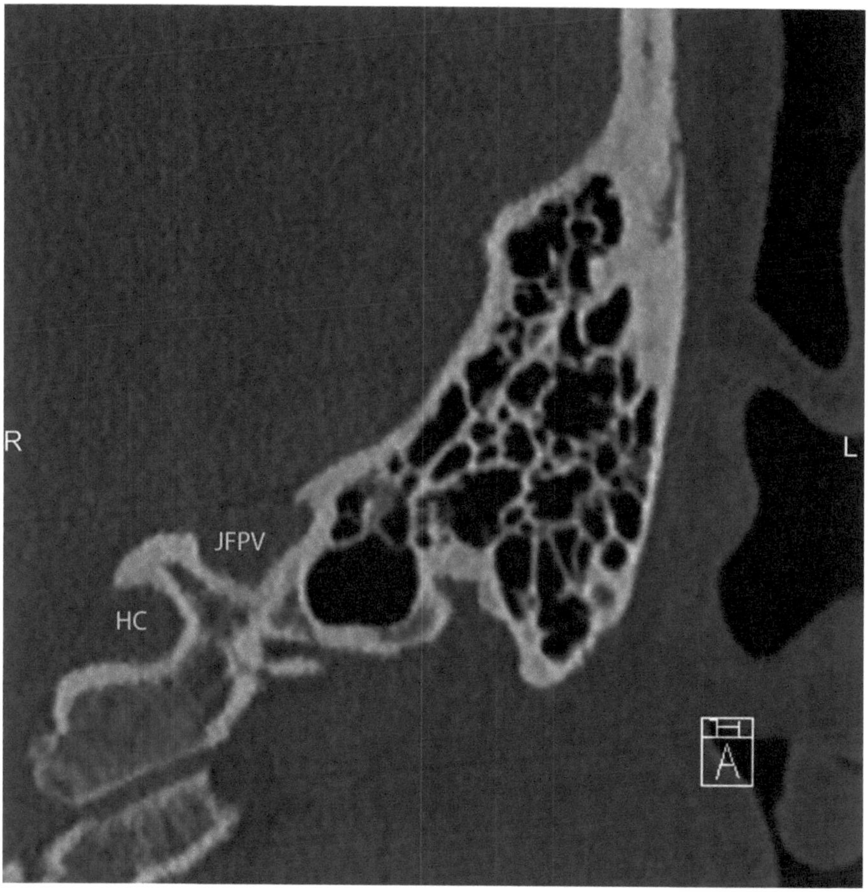

C48: HRCT coronal reformat

HC hypoglossal canal
JFPV jugular foramen – pars vascularis

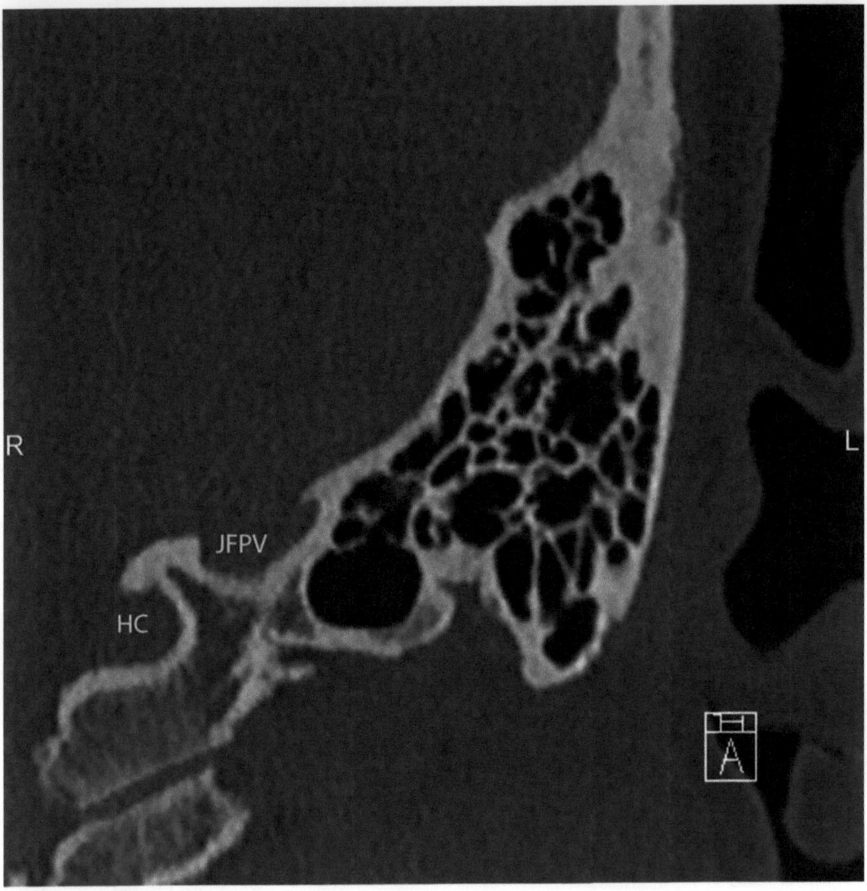

C49: HRCT coronal reformat

HC hypoglossal canal
JFPV jugular foramen – pars vascularis

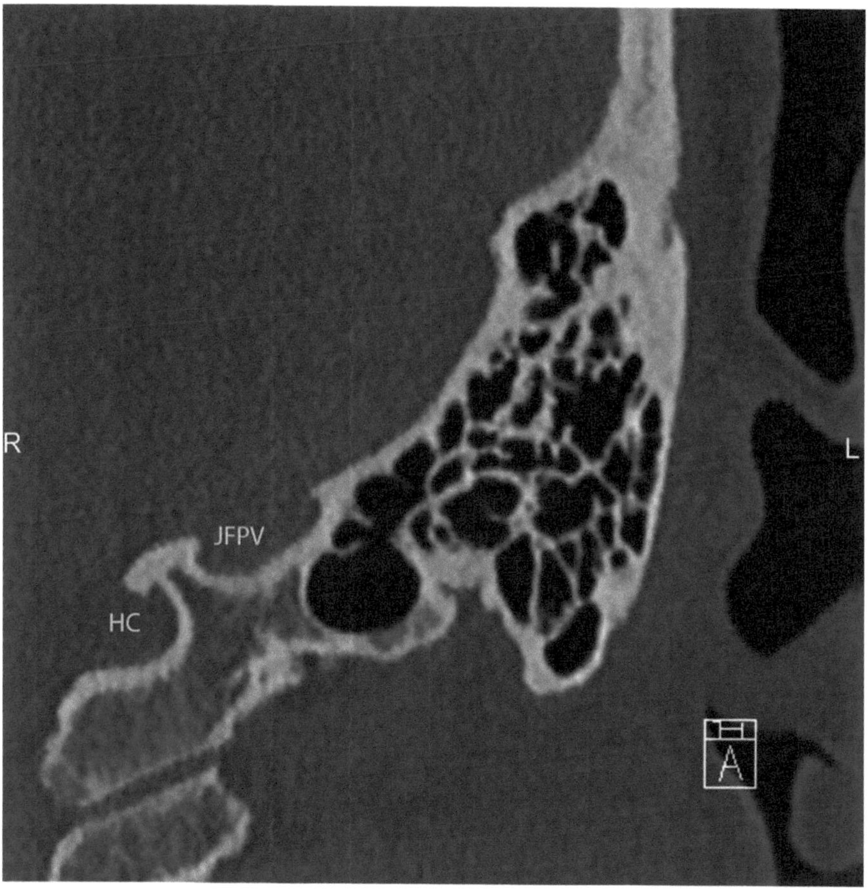

C50: HRCT coronal reformat

HC	hypoglossal canal
JFPV	jugular foramen – pars vascularis

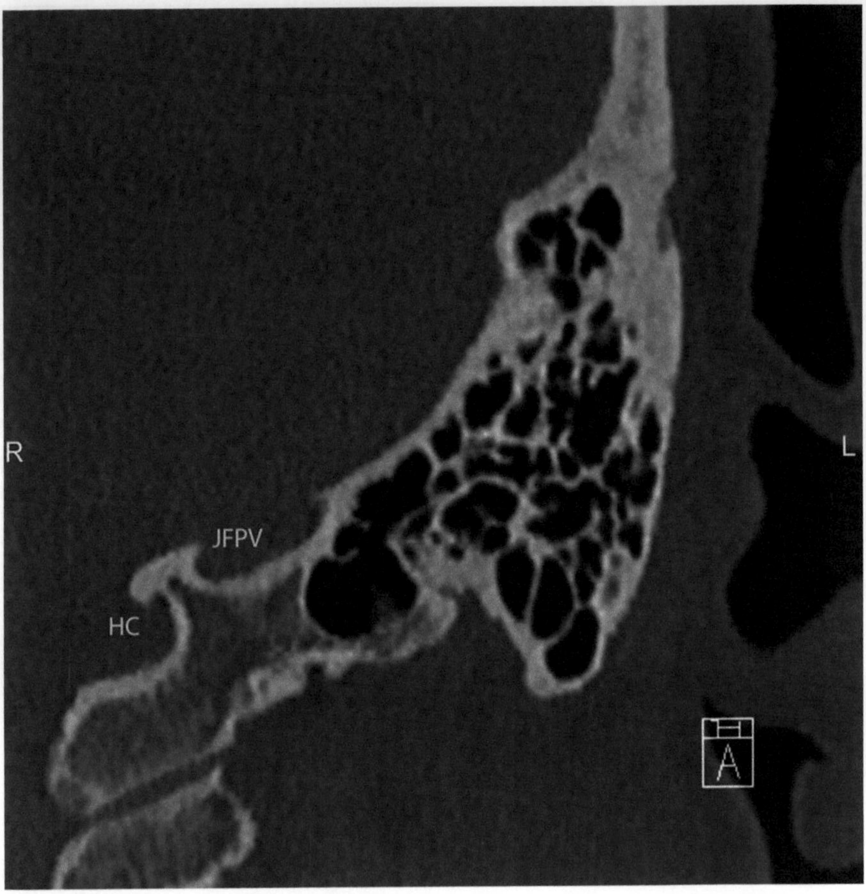

C51: HRCT coronal reformat

HC	hypoglossal canal
JFPV	jugular foramen – pars vascularis

Chapter 4
HRCT Stenvers Reformat

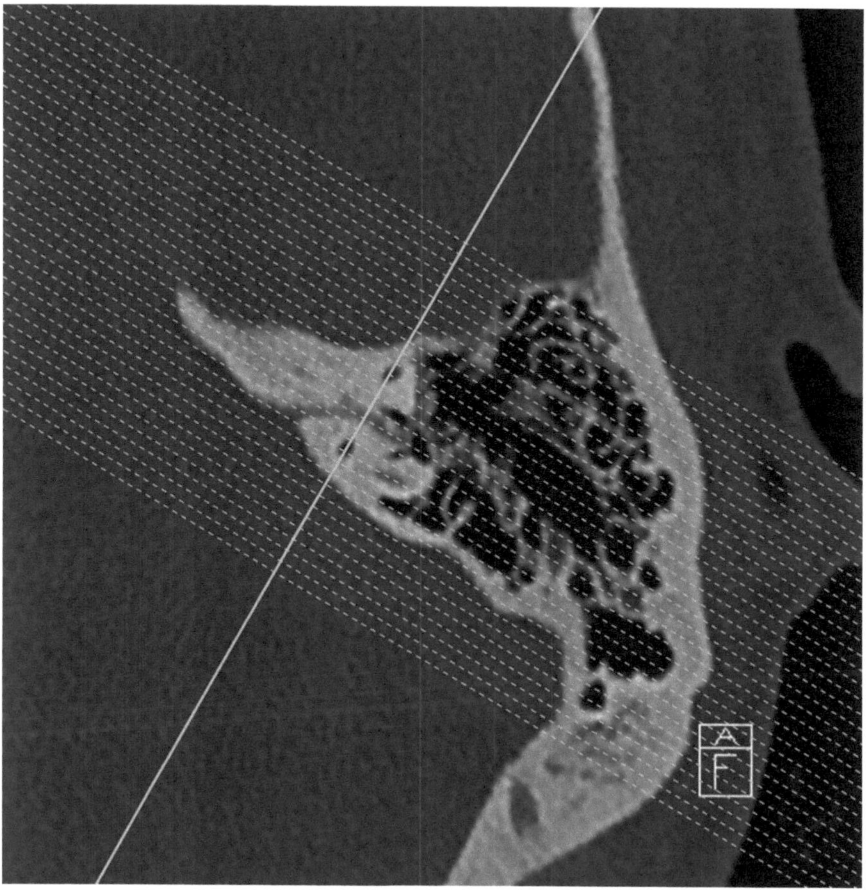

S00: HRCT Stenvers reformat

Stenvers plane (oblique coronal, i.e. long axis of the pyramid)

Using an MPR 3D viewer, anterior semicircular canal is identified on reformatted axial sections. The Stenvers reformat runs perpendicular to the plane defined by the anterior semicircular canal.

J. Kopřiva, J. Žižka, *Temporal Bone CT and MRI Anatomy: A Guide to 3D Volumetric Acquisitions*, DOI 10.1007/978-3-319-08242-4_4

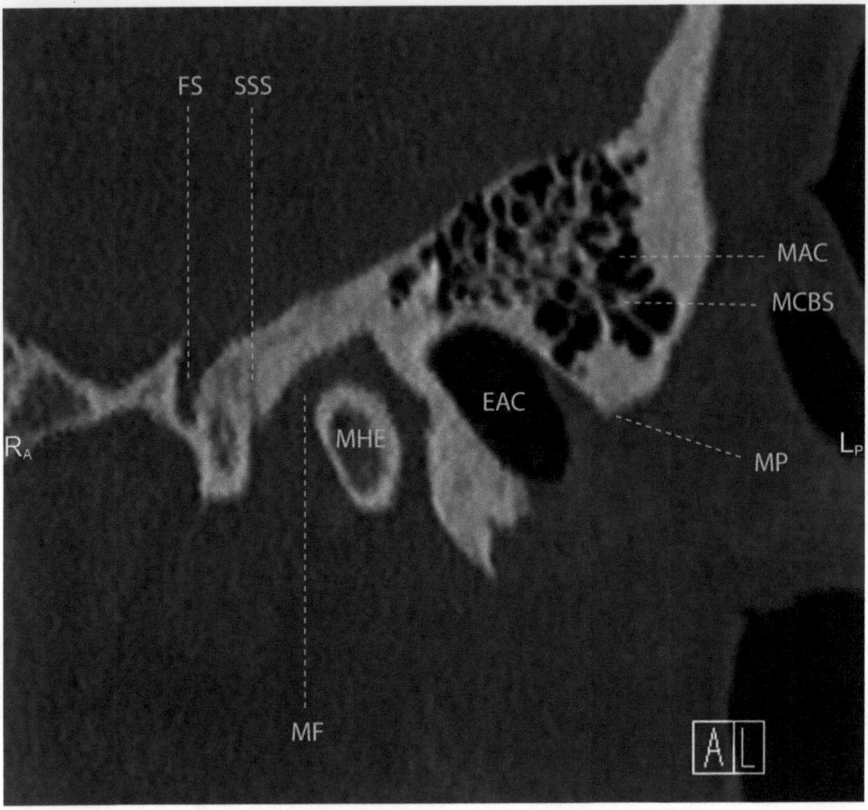

S01: HRCT Stenvers reformat

EAC	external auditory canal	MF	mandibular fossa
FS	foramen spinosum	MHE	mandible – head
MAC	mastoid air cells	MP	mastoid process
MCBS	mastoid cells bony septa	SSS	sphenosquamosal suture

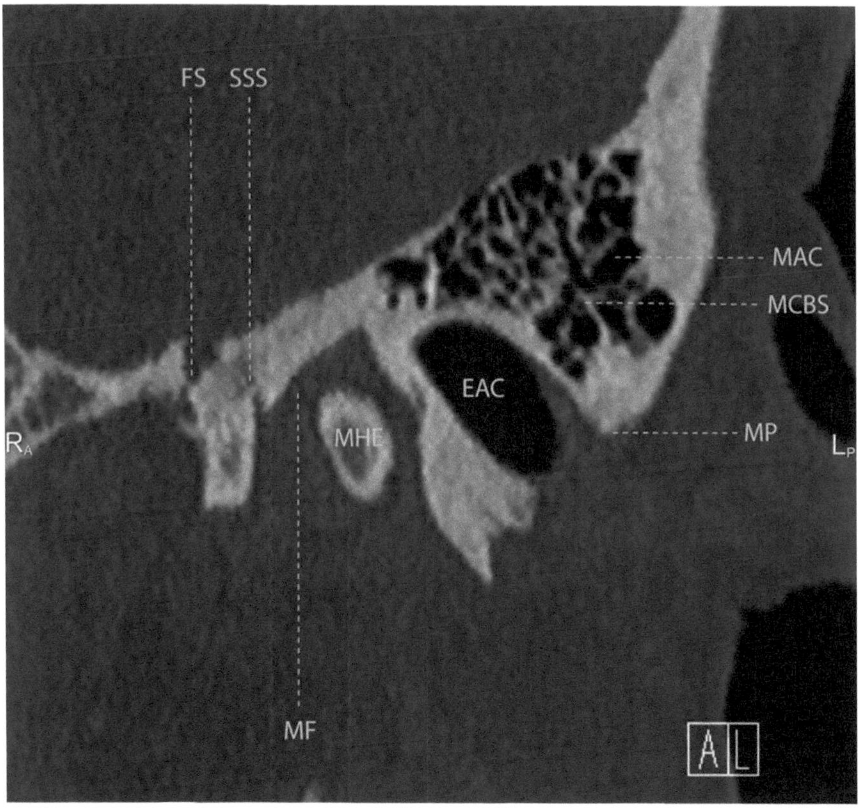

S02: HRCT Stenvers reformat

EAC	external auditory canal	MF	mandibular fossa
FS	foramen spinosum	MHE	mandible – head
MAC	mastoid air cells	MP	mastoid process
MCBS	mastoid cells bony septa	SSS	sphenosquamosal suture

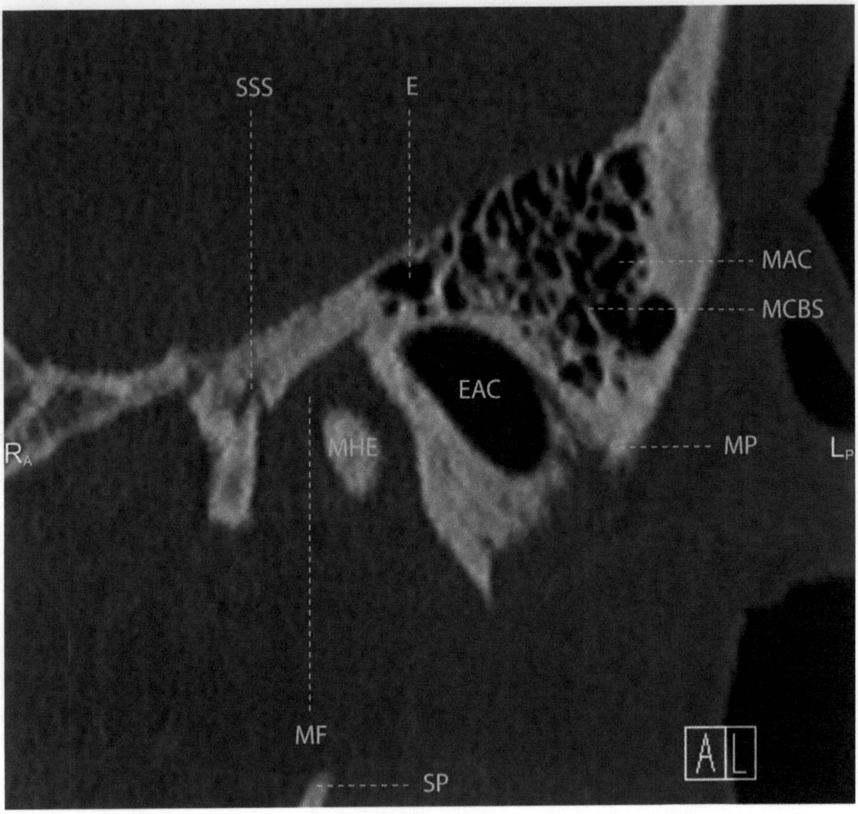

S03: HRCT Stenvers reformat

E	epitympanum	MHE	mandible – head
EAC	external auditory canal	MP	mastoid process
MAC	mastoid air cells	SP	styloid process
MCBS	mastoid cells bony septa	SSS	sphenosquamosal suture
MF	mandibular fossa		

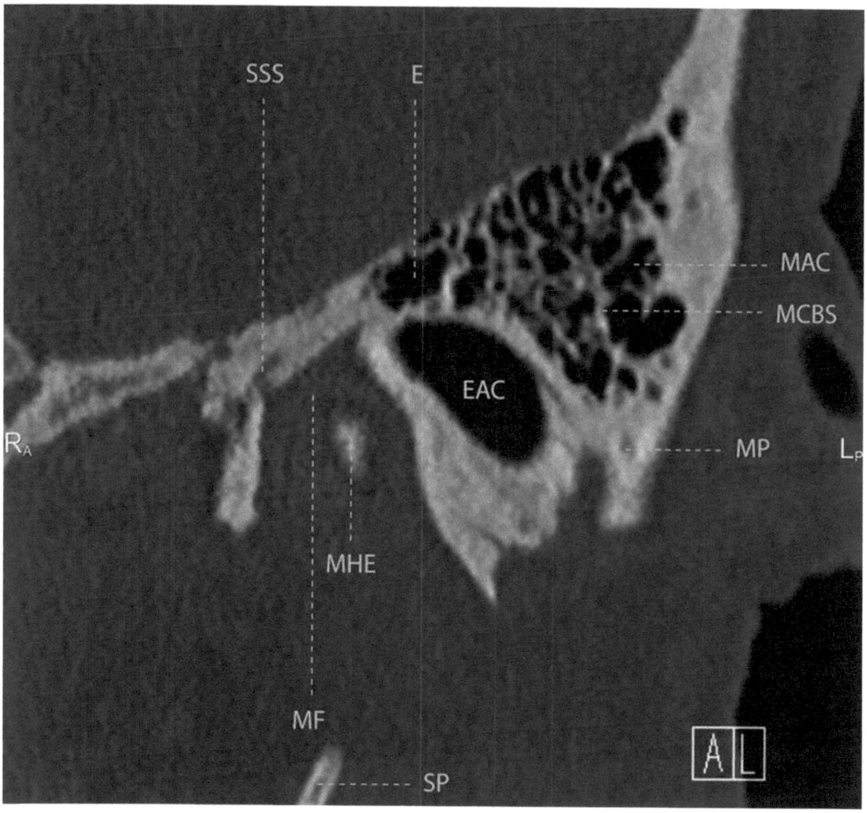

S04: HRCT Stenvers reformat

E	epitympanum
EAC	external auditory canal
MAC	mastoid air cells
MCBS	mastoid cells bony septa
MF	mandibular fossa
MHE	mandible – head
MP	mastoid process
SP	styloid process
SSS	sphenosquamosal suture

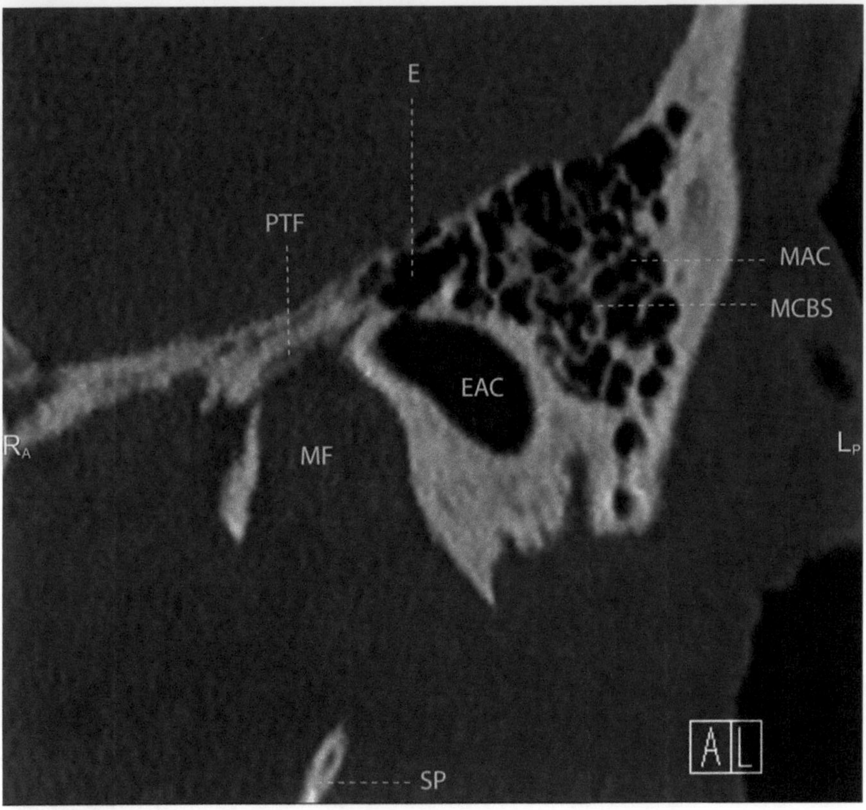

S05: HRCT Stenvers reformat

E	epitympanum
EAC	external auditory canal
MAC	mastoid air cells
MCBS	mastoid cells bony septa
MF	mandibular fossa
PTF	petrotympanic fissure
SP	styloid process

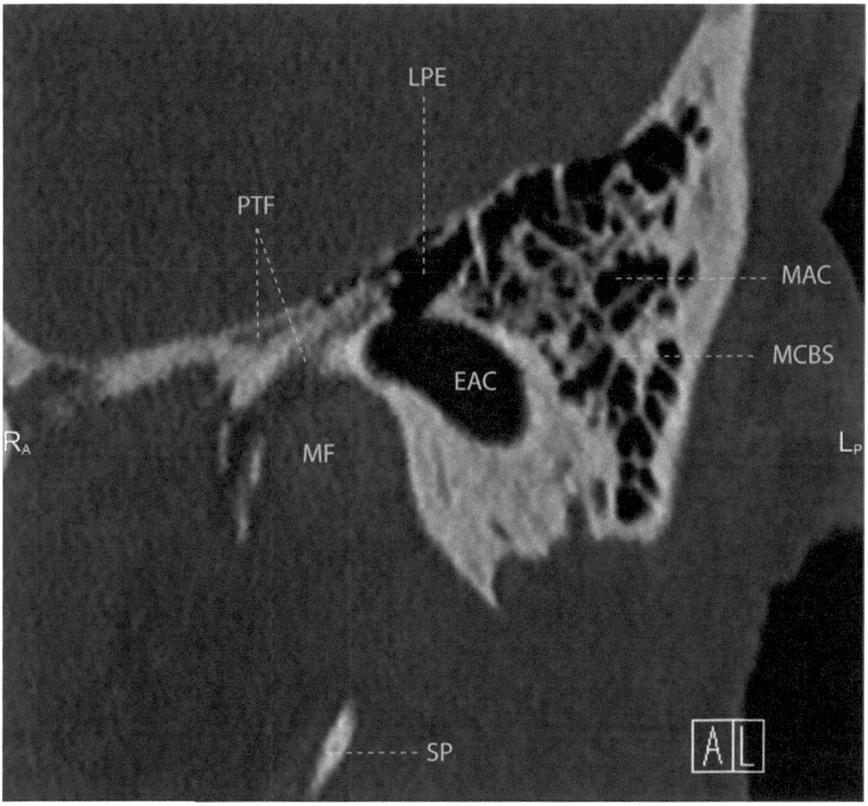

S06: HRCT Stenvers reformat

EAC	external auditory canal	MF	mandibular fossa
LPE	lateral posterior epitympanum	PTF	petrotympanic fissure
MAC	mastoid air cells	SP	styloid process
MCBS	mastoid cells bony septa		

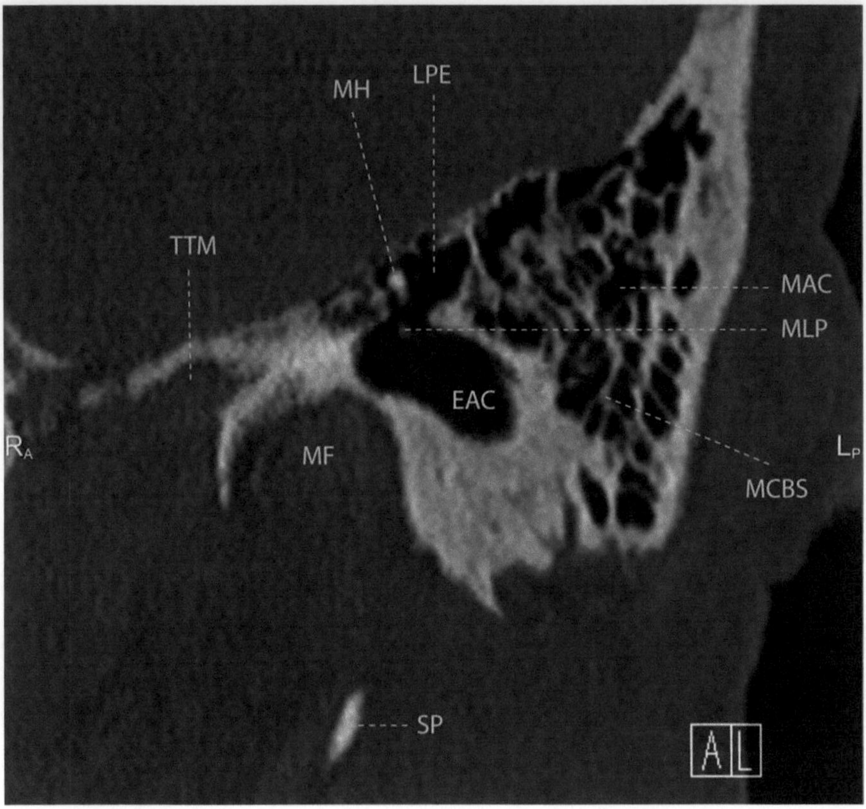

S07: HRCT Stenvers reformat

EAC	external auditory canal	MH	malleus – head
LPE	lateral posterior epitympanum	MLP	malleus – lateral process
MAC	mastoid air cells	SP	styloid process
MCBS	mastoid cells bony septa	TTM	tensor tympani muscle
MF	mandibular fossa		

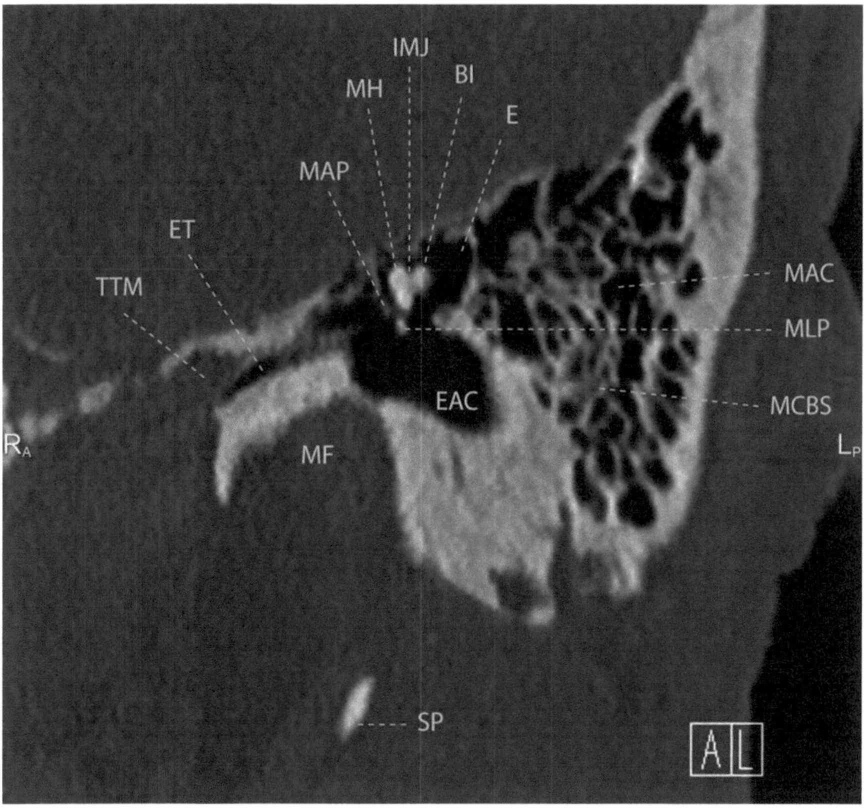

S08: HRCT Stenvers reformat

BI	body of incus	MCBS	mastoid cells bony septa
E	epitympanum	MF	mandibular fossa
EAC	external auditory canal	MH	malleus – head
ET	Eustachian tube	MLP	malleus – lateral process
IMJ	incudomallear joint	SP	styloid process
MAC	mastoid air cells	TTM	tensor tympani muscle
MAP	malleus – anterior process		

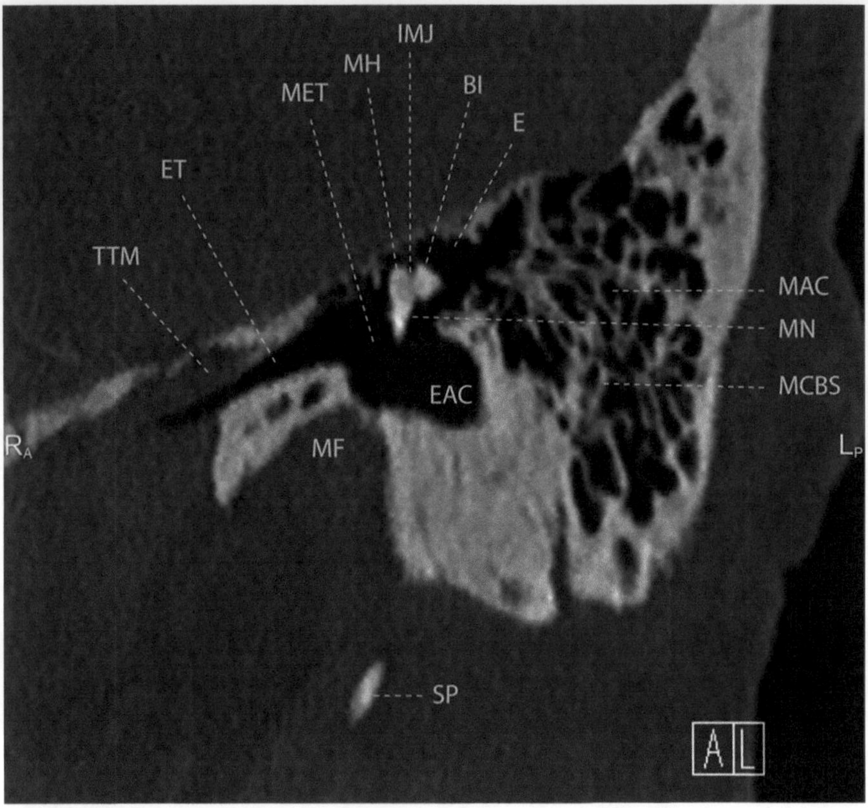

S09: HRCT Stenvers reformat

BI	body of incus	MET	mesotympanum
E	epitympanum	MF	mandibular fossa
EAC	external auditory canal	MH	malleus – head
ET	Eustachian tube	MN	malleus – neck
IMJ	incudomallear joint	SP	styloid process
MAC	mastoid air cells	TTM	tensor tympani muscle
MCBS	mastoid cells bony septa		

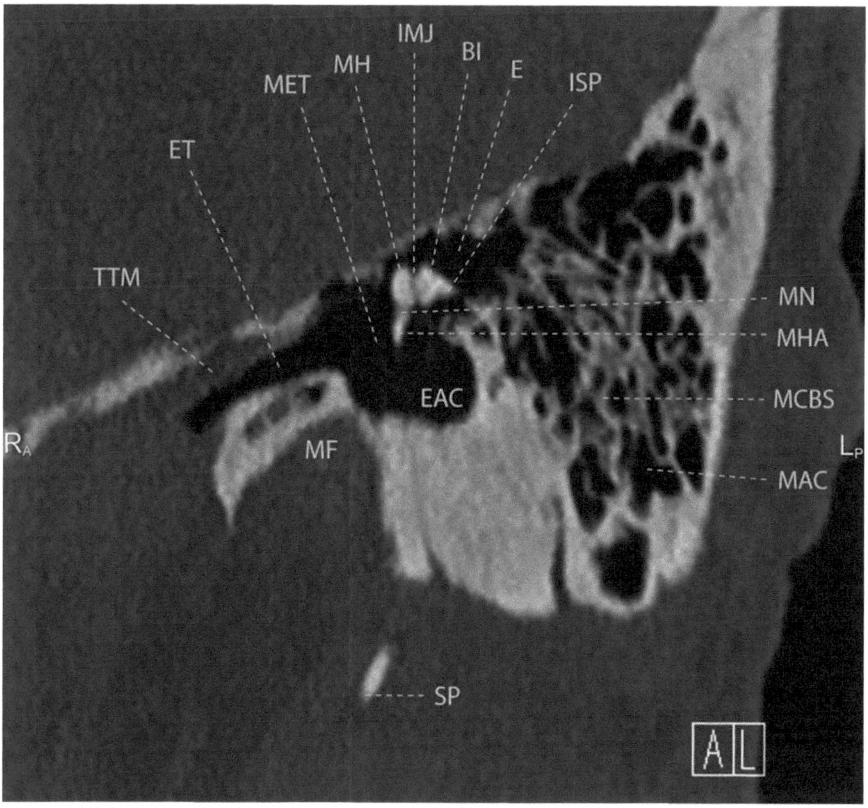

S10: HRCT Stenvers reformat

BI	body of incus	MET	mesotympanum
E	epitympanum	MF	mandibular fossa
EAC	external auditory canal	MH	malleus – head
ET	Eustachian tube	MHA	malleus – handle
IMJ	incudomallear joint	MN	malleus – neck
ISP	incus – short process	SP	styloid process
MAC	mastoid air cells	TTM	tensor tympani muscle
MCBS	mastoid cells bony septa		

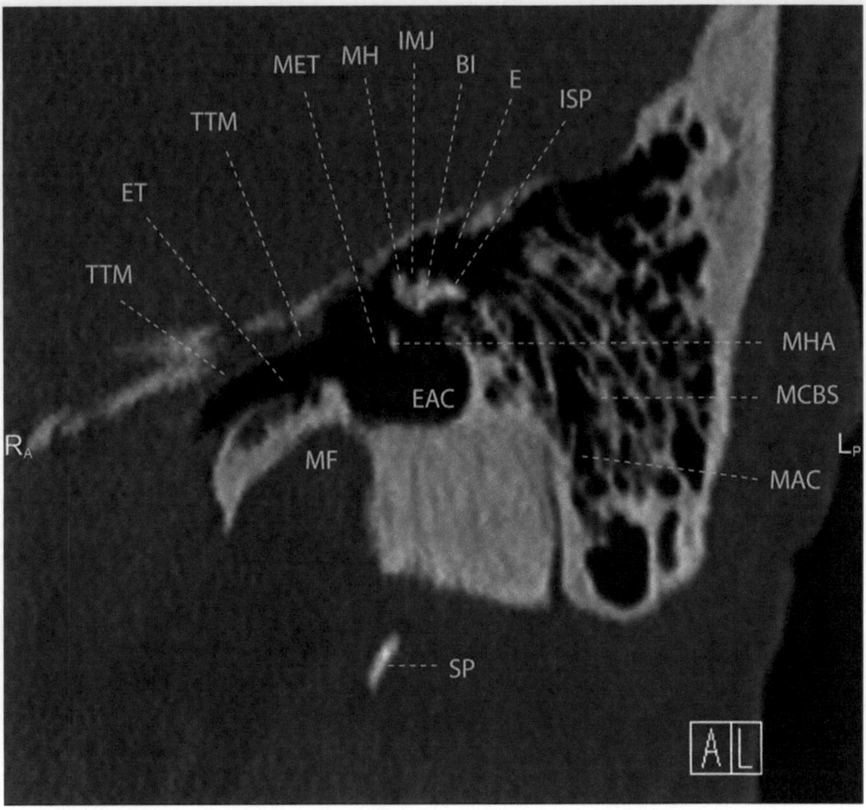

S11: HRCT Stenvers reformat

BI	body of incus	MCBS	mastoid cells bony septa
E	epitympanum	MET	mesotympanum
EAC	external auditory canal	MF	mandibular fossa
ET	Eustachian tube	MH	malleus – head
IMJ	incudomallear joint	MHA	malleus – handle
ISP	incus – short process	SP	styloid process
MAC	mastoid air cells	TTM	tensor tympani muscle

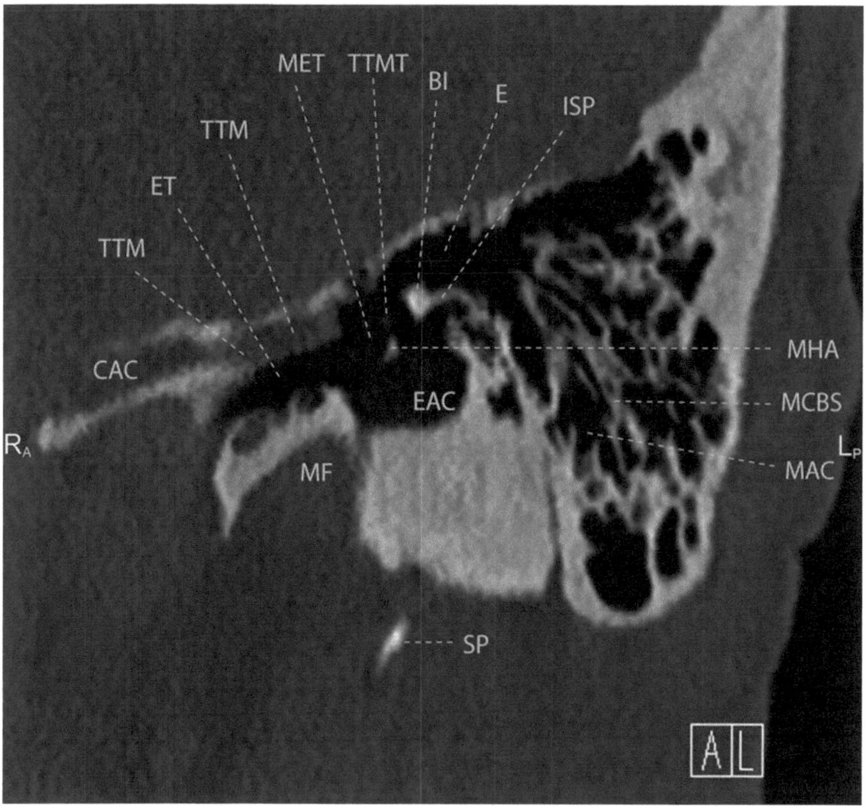

S12: HRCT Stenvers reformat

BI	body of incus	MET	mesotympanum
CAC	carotid canal	MF	mandibular fossa
E	epitympanum	MHA	malleus – handle
EAC	external auditory canal	SP	styloid process
ET	Eustachian tube	TTM	tensor tympani muscle
ISP	incus – short process	TTMT	tensor tympani muscle – tendon
MAC	mastoid air cells		
MCBS	mastoid cells bony septa		

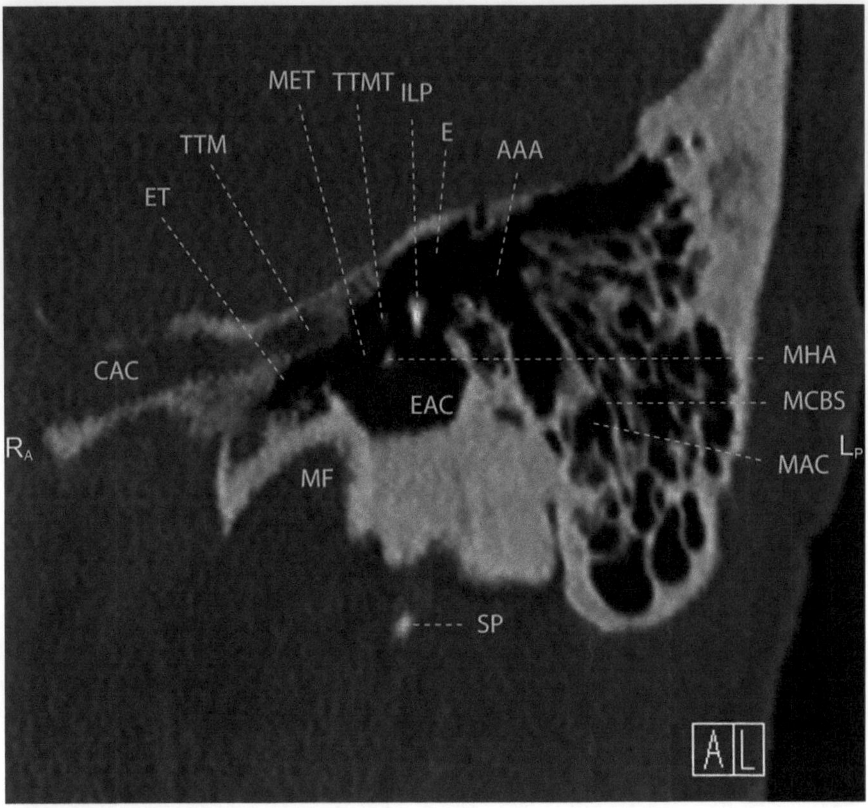

S13: HRCT Stenvers reformat

AAA	aditus ad antrum	MET	mesotympanum
CAC	carotid canal	MF	mandibular fossa
E	epitympanum	MHA	malleus – handle
EAC	external auditory canal	SP	styloid process
ET	Eustachian tube	TTM	tensor tympani muscle
ILP	incus – long process	TTMT	tensor tympani muscle – tendon
MAC	mastoid air cells		
MCBS	mastoid cells bony septa		

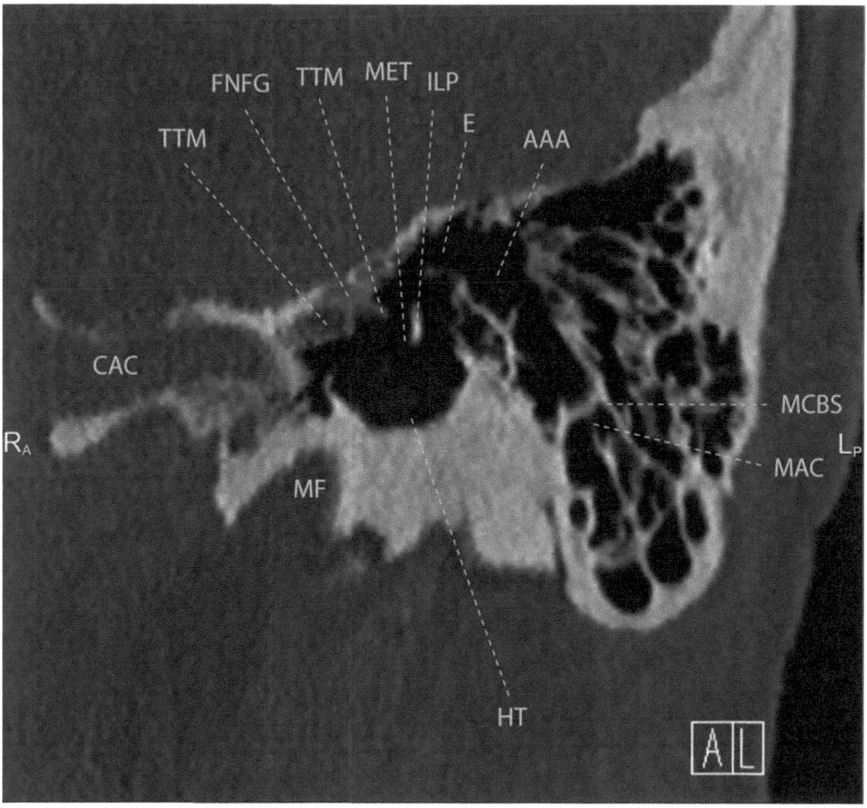

S14: HRCT Stenvers reformat

AAA	aditus ad antrum	MAC	mastoid air cells
CAC	carotid canal	MCBS	mastoid cells bony septa
E	epitympanum	MET	mesotympanum
FNFG	facial nerve canal – first genu	MF	mandibular fossa
HT	hypotympanum	TTM	tensor tympani muscle
ILP	incus – long process		

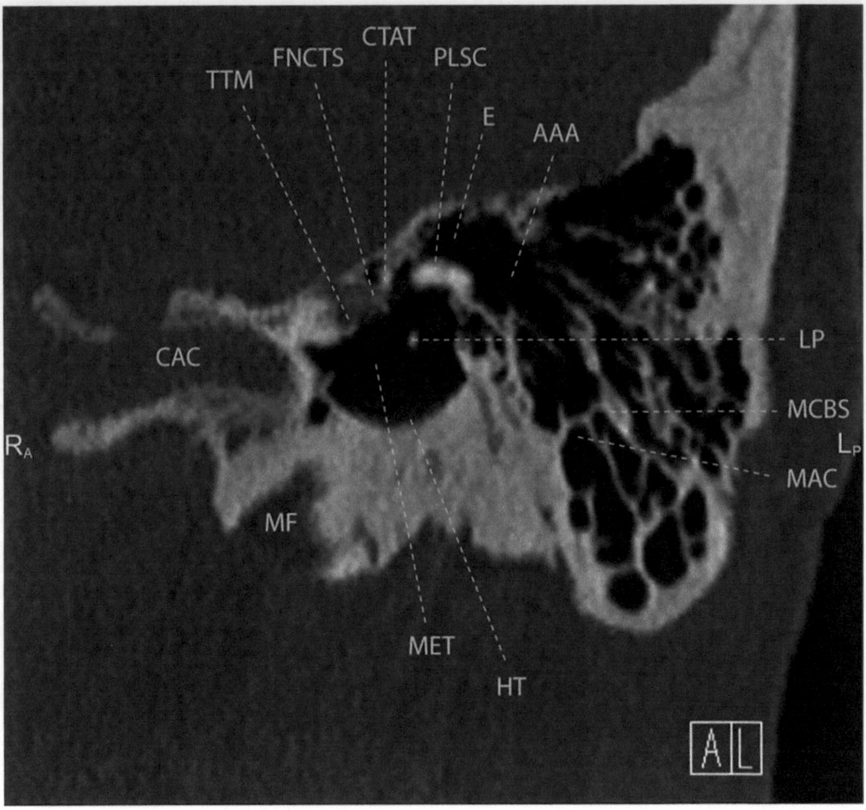

S15: HRCT Stenvers reformat

AAA	aditus ad antrum	LP	lenticular process
CAC	carotid canal	MAC	mastoid air cells
CTAT	crista transversa – anterior epitympanum	MCBS	mastoid cells bony septa
		MET	mesotympanum
E	epitympanum	MF	mandibular fossa
FNCTS	facial nerve canal – tympanic segment	PLSC	prominence of lateral semicircular canal
HT	hypotympanum	TTM	tensor tympani muscle

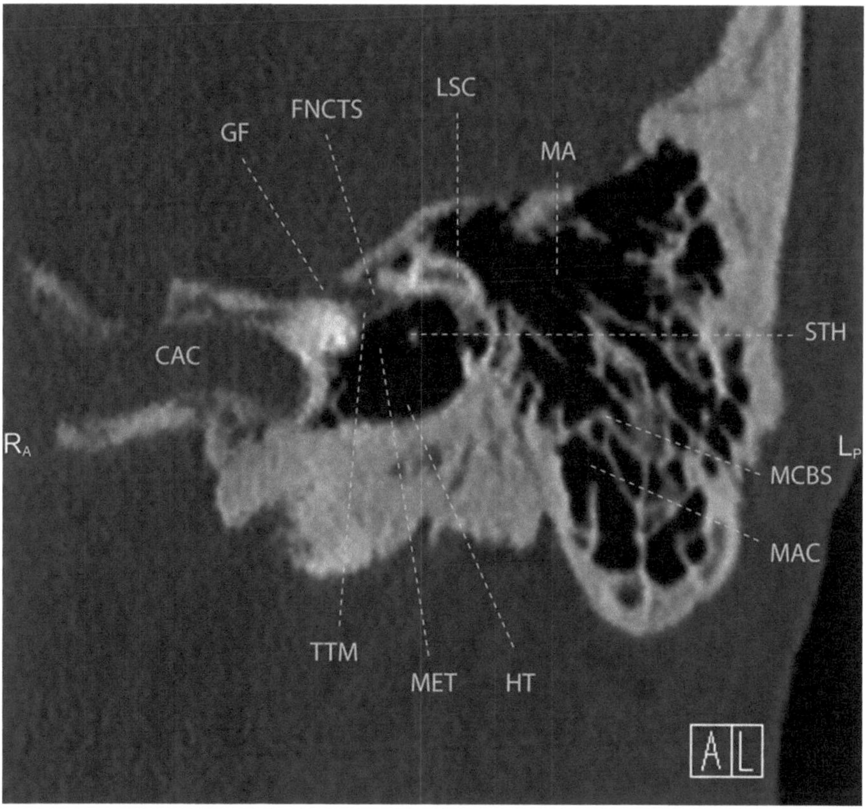

S16: HRCT Stenvers reformat

CAC	carotid canal
FNCTS	facial nerve canal tympanic segment
GF	geniculate fossa
HT	hypotympanum
LSC	lateral semicircular canal
MA	mastoid antrum
MAC	mastoid air cells
MCBS	mastoid cells bony septa
MET	mesotympanum
STH	stapes – head
TTM	tensor tympani muscle

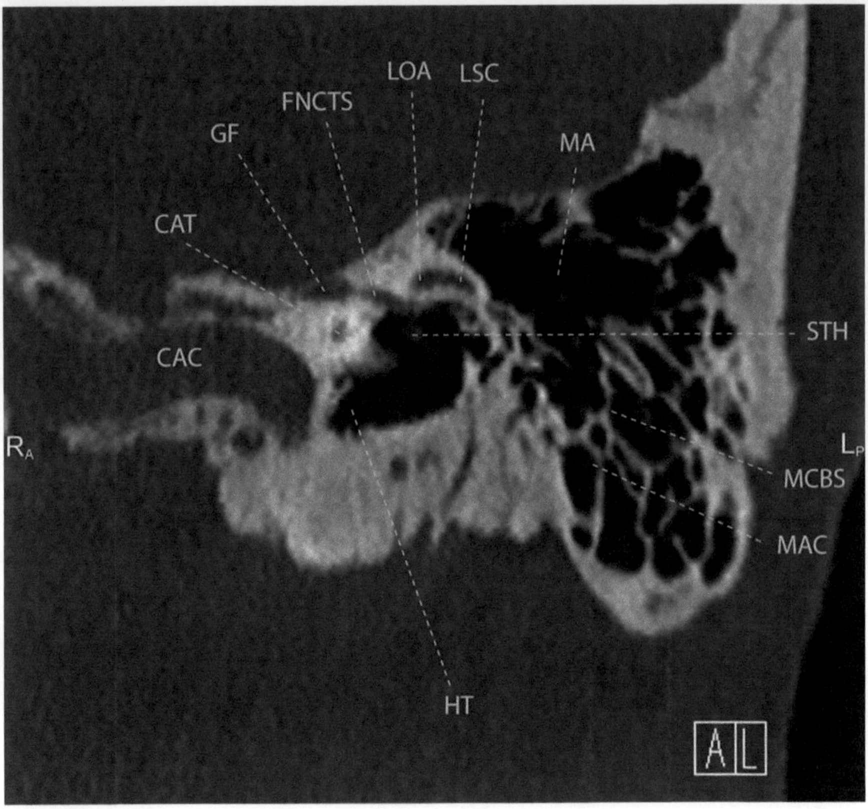

S17: HRCT Stenvers reformat

CAT	cochlea – apical turn	LOA	lateral osseous ampulla
CAC	carotid canal	LSC	lateral semicircular canal
FNCTS	facial nerve canal tympanic segment	MA	mastoid antrum
		MAC	mastoid air cells
GF	geniculate fossa	MCBS	mastoid cells bony septa
HT	hypotympanum	STH	stapes – head

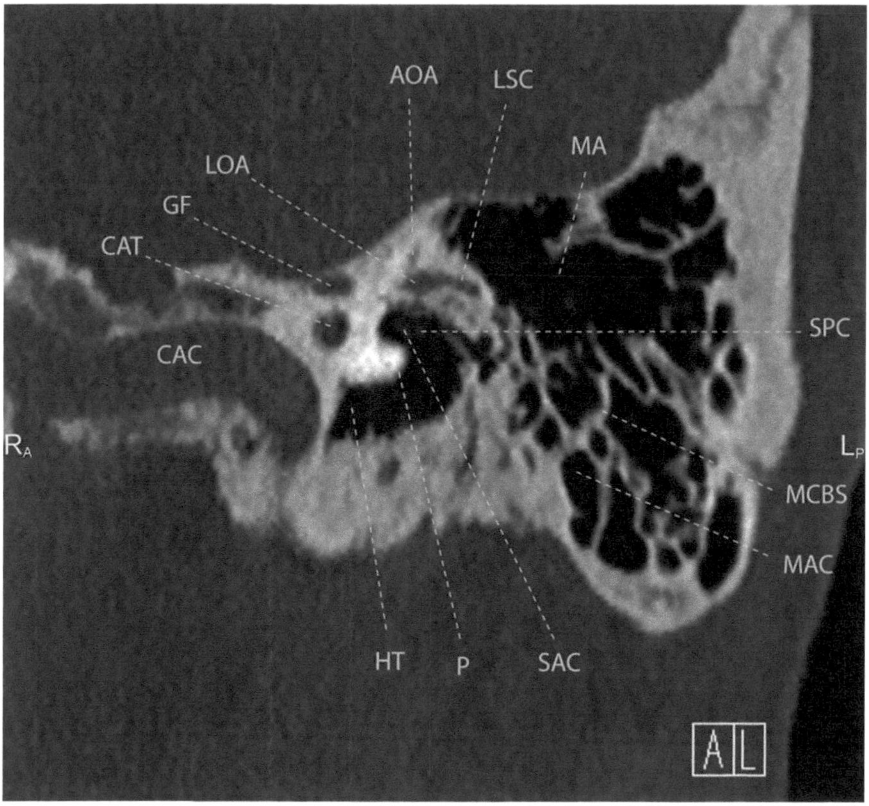

S18: HRCT Stenvers reformat

AOA	anterior osseous ampulla	MA	mastoid antrum
CAC	carotid canal	MAC	mastoid air cells
CAT	cochlea – apical turn	MCBS	mastoid cells bony septa
GF	geniculate fossa	P	promontory
HT	hypotympanum	SAC	stapes – anterior crus
LOA	lateral osseous ampulla	SPC	stapes – posterior crus
LSC	lateral semicircular canal		

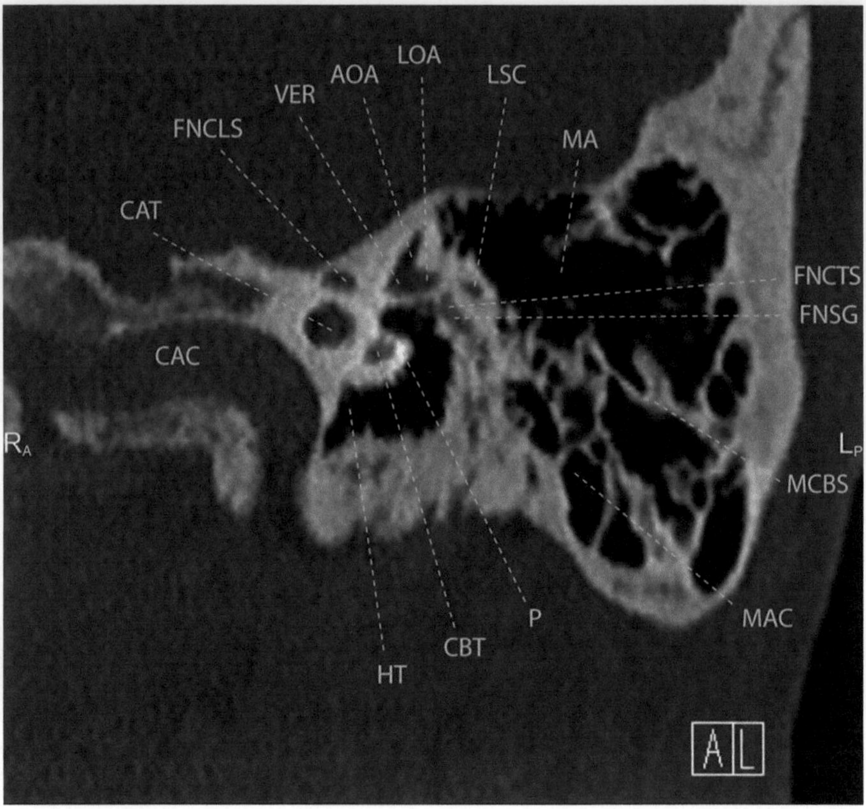

S19: HRCT Stenvers reformat

AOA	anterior osseous ampulla	HT	hypotympanum
CAC	carotid canal	LOA	lateral osseous ampulla
CAT	cochlea – apical turn	LSC	lateral semicircular canal
CBT	cochlea – basilar turn	MA	mastoid antrum
FNCLS	facial nerve canal – labyrinthine segment	MAC	mastoid air cells
FNCTS	facial nerve canal tympanic segment	MCBS	mastoid cells bony septa
FNSG	facial nerve canal – second genu	P	promontory
		VER	vestibule – elliptical recess

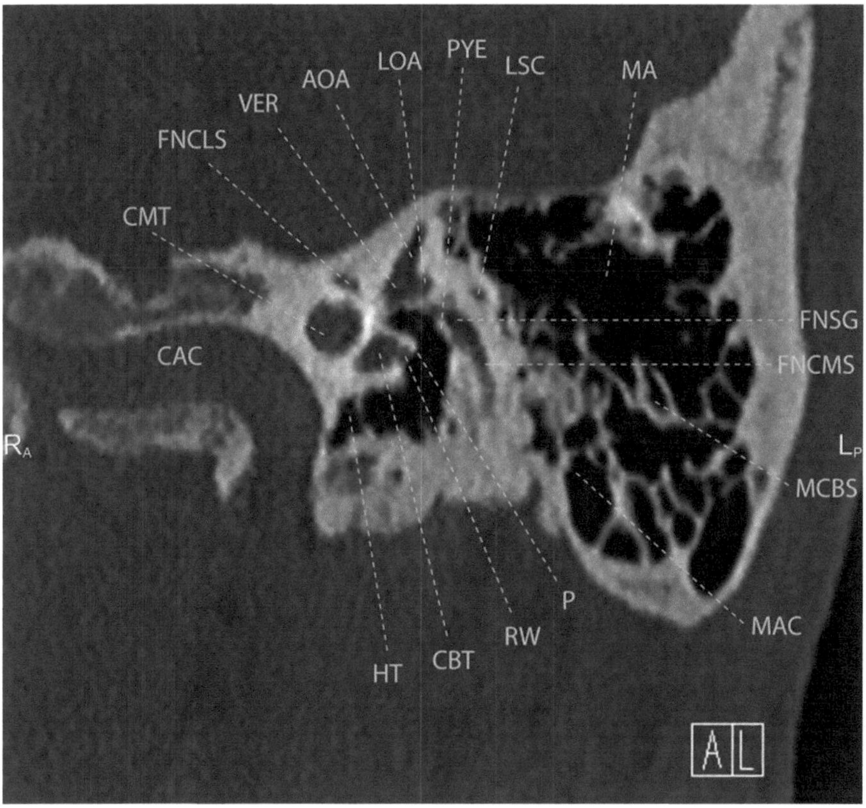

S20: HRCT Stenvers reformat

AOA	anterior osseous ampulla	HT	hypotympanum
CAC	carotid canal	LOA	lateral osseous ampulla
CBT	cochlea basilar turn	LSC	lateral semicircular canal
CMT	cochlea – middle turn	MA	mastoid antrum
FNCLS	facial nerve canal – labyrinthine segment	MAC	mastoid air cells
		MCBS	mastoid cells bony septa
FNCMS	facial nerve canal – mastoid segment	P	promontory
		PYE	pyramidal eminence
FNSG	facial nerve canal – second genu	RW	round window
		VER	vestibule – elliptical recess

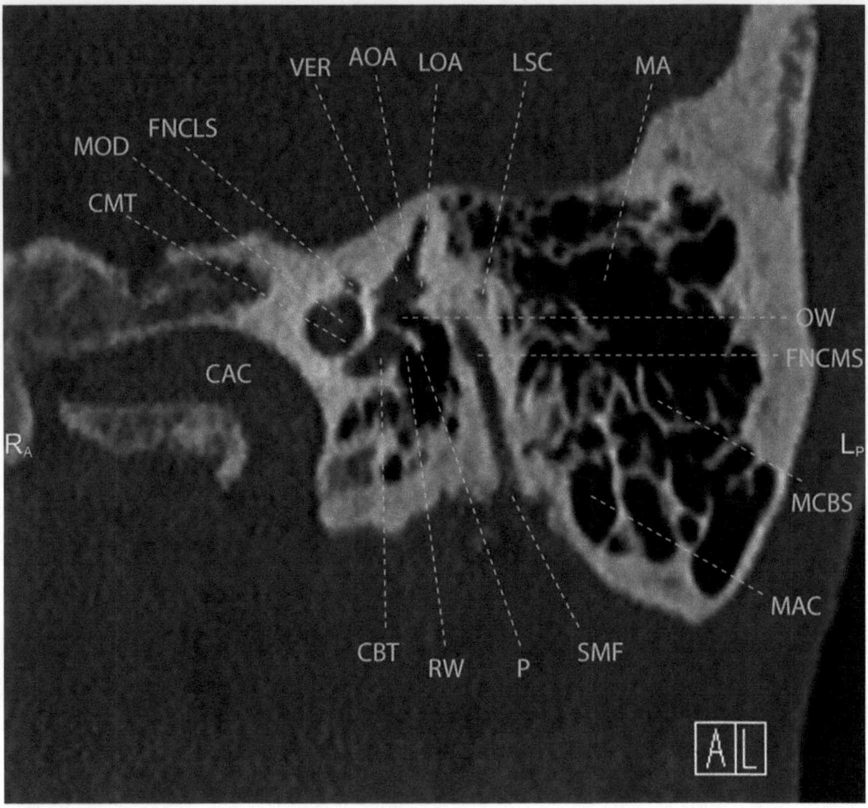

S21: HRCT Stenvers reformat

AOA	anterior osseous ampulla	MA	mastoid antrum
CAC	carotid canal	MAC	mastoid air cells
CBT	cochlea – basilar turn	MCBS	mastoid cells bony septa
CMT	cochlea – middle turn	MOD	modiolus
FNCLS	facial nerve canal – labyrinthine segment	OW	oval window
		P	promontory
FNCMS	facial nerve canal – mastoid segment	RW	round window
		SMF	stylomastoid foramen
LOA	lateral osseous ampulla	VER	vestibule – elliptical recess
LSC	lateral semicircular canal		

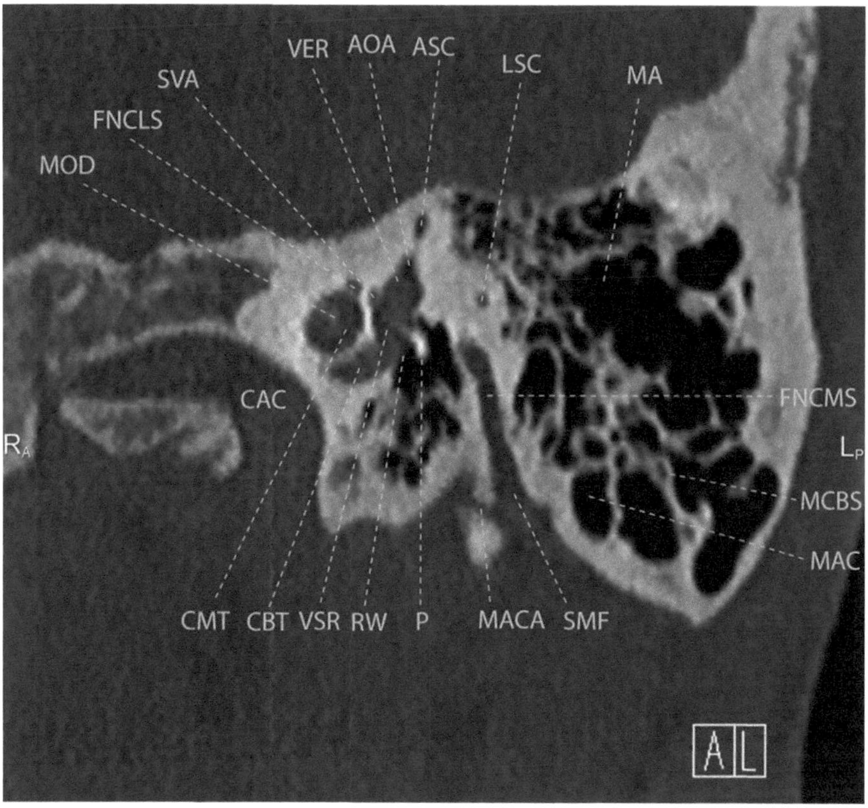

S22: HRCT Stenvers reformat

AOA	anterior osseous ampulla
ASC	anterior semicircular canal
CAC	carotid canal
CBT	cochlea basilar turn
CMT	cochlea – middle turn
FNCLS	facial nerve canal – labyrinthine segment
FNCMS	facial nerve canal – mastoid segment
LSC	lateral semicircular canal
MA	mastoid antrum
MAC	mastoid air cells
MACA	mastoid canaliculus
MCBS	mastoid cells bony septa
MOD	modiolus
P	promontory
RW	round window
SMF	stylomastoid foramen
SVA	superior vestibular area
VER	vestibule – elliptical recess
VSR	vestibule – spherical recess

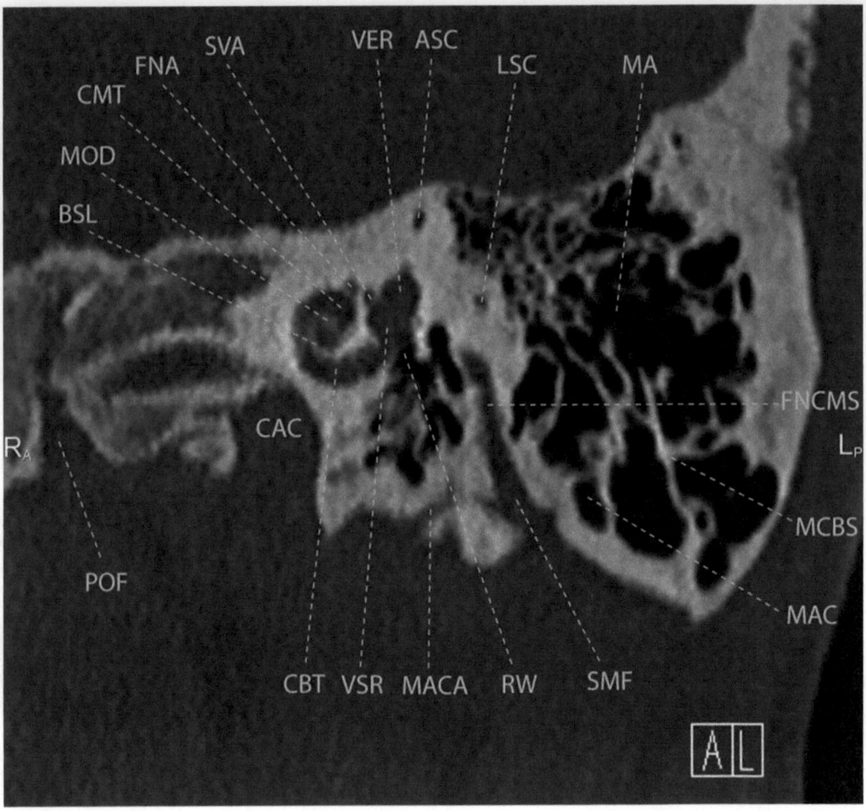

S23: HRCT Stenvers reformat

ASC	anterior semicircular canal	MAC	mastoid air cells
BSL	bony spiral lamina	MACA	mastoid canaliculus
CAC	carotid canal	MCBS	mastoid cells bony septa
CBT	cochlea – basilar turn	MOD	modiolus
CMT	cochlea – middle turn	POF	petrooccipital fissure
FNA	facial nerve area	RW	round window
FNCMS	facial nerve canal – mastoid segment	SMF	stylomastoid foramen
LSC	lateral semicircular canal	SVA	superior vestibular area
MA	mastoid antrum	VER	vestibule – elliptical recess
		VSR	vestibule – spherical recess

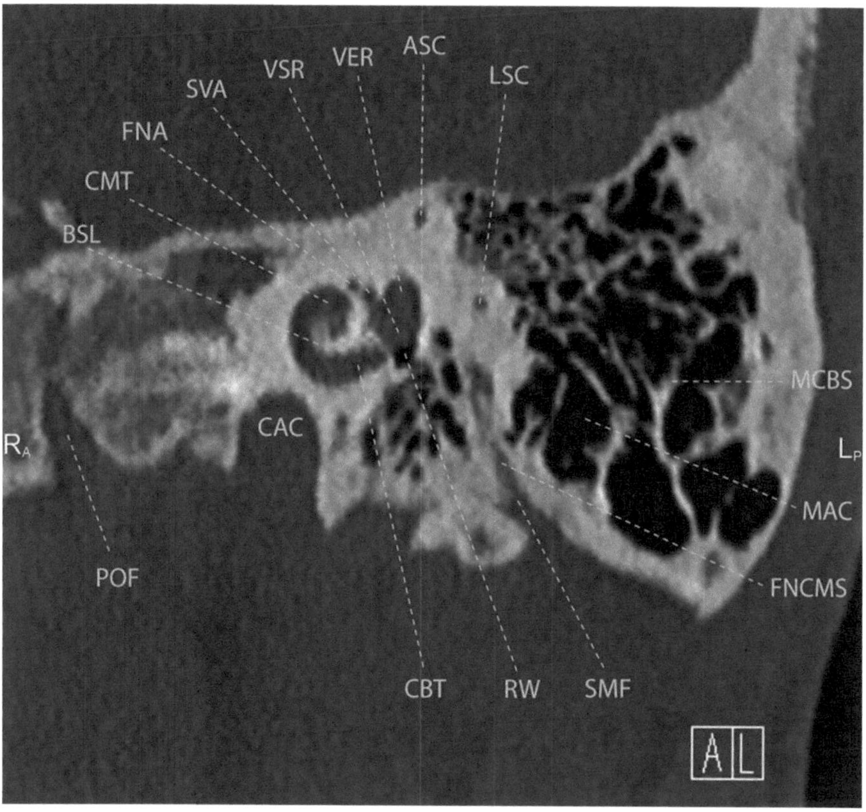

S24: HRCT Stenvers reformat

ASC	anterior semicircular canal
BSL	bony spiral lamina
CAC	carotid canal
CBT	cochlea – basilar turn
CMT	cochlea – middle turn
FNA	facial nerve area
FNCMS	facial nerve canal – mastoid segment
LSC	lateral semicircular canal
MAC	mastoid air cells
MCBS	mastoid cells bony septa
POF	petrooccipital fissure
RW	round window
SMF	stylomastoid foramen
SVA	superior vestibular area
VER	vestibule – elliptical recess
VSR	vestibule – spherical recess

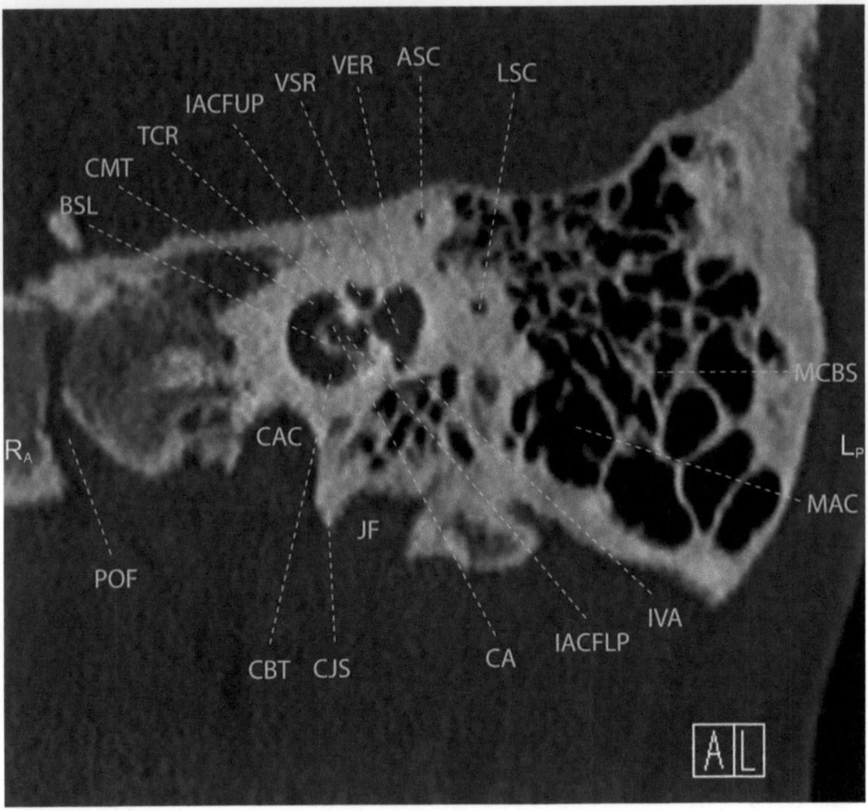

S25: HRCT Stenvers reformat

ASC	anterior semicircular canal
BSL	bony spiral lamina
CA	cochlear area
CAC	carotid canal
CBT	cochlea – basilar turn
CJS	caroticojugular spine
CMT	cochlea – middle turn
IACFLP	internal auditory canal – fundus lower part
IACFUP	internal auditory canal – fundus upper part
IVA	inferior vestibular area
JF	jugular foramen
LSC	lateral semicircular canal
MAC	mastoid air cells
MCBS	mastoid cells bony septa
POF	petrooccipital fissure
TCR	transverse crest
VER	vestibule – elliptical recess
VSR	vestibule – spherical recess

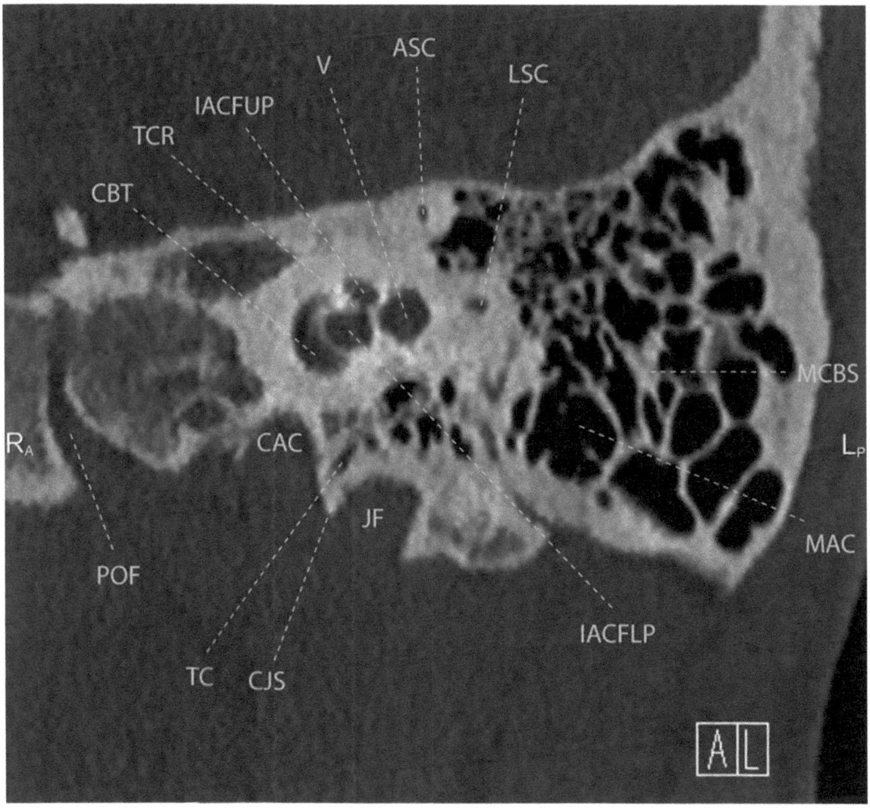

S26: HRCT Stenvers reformat

ASC	anterior semicircular canal	JF	jugular foramen
CAC	carotid canal	LSC	lateral semicircular canal
CBT	cochlea – basilar turn	MAC	mastoid air cells
CJS	caroticojugular spine	MCBS	mastoid cells bony septa
IACFLP	internal auditory canal – fundus lower part	POF	petrooccipital fissure
		TC	tympanic canaliculus
IACFUP	internal auditory canal – fundus upper part	TCR	transverse crest
		V	vestibule

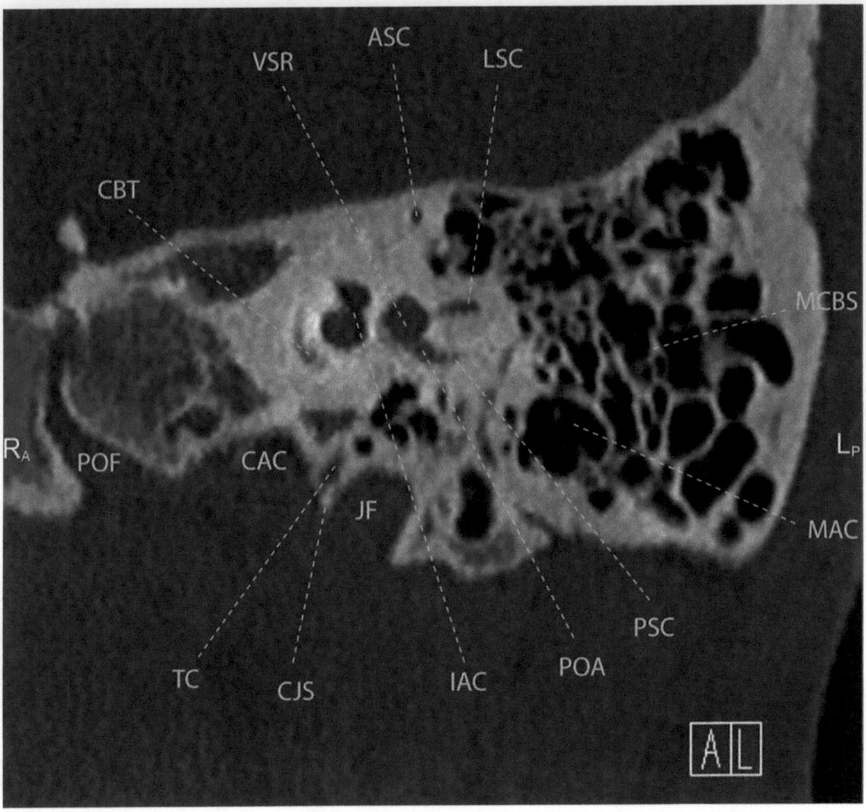

S27: HRCT Stenvers reformat

ASC	anterior semicircular canal	MAC	mastoid air cells
CAC	carotid canal	MCBS	mastoid cells bony septa
CBT	cochlea – basilar turn	POA	posterior osseous ampulla
CJS	caroticojugular spine	POF	petrooccipital fissure
IAC	internal auditory canal	PSC	posterior semicircular canal
JF	jugular foramen	TC	tympanic canaliculus
LSC	lateral semicircular canal	VSR	vestibule – spherical recess

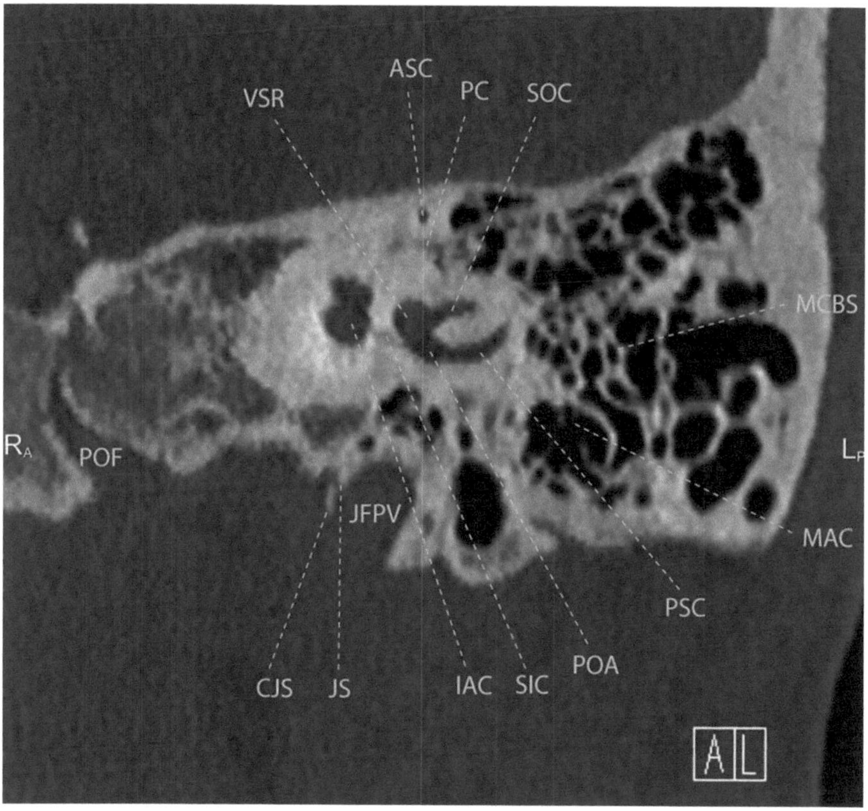

S28: HRCT Stenvers reformat

ASC	anterior semicircular canal	PC	petromastoid canal
CJS	caroticojugular spine	POA	posterior osseous ampulla
IAC	internal auditory canal	POF	petrooccipital fissure
JFPV	jugular foramen – pars vascularis	PSC	posterior semicircular canal
		SIC	singular canal
JS	jugular spine	SOC	simple osseous crus
MAC	mastoid air cells	VSR	vestibule – spherical recess
MCBS	mastoid cells bony septa		

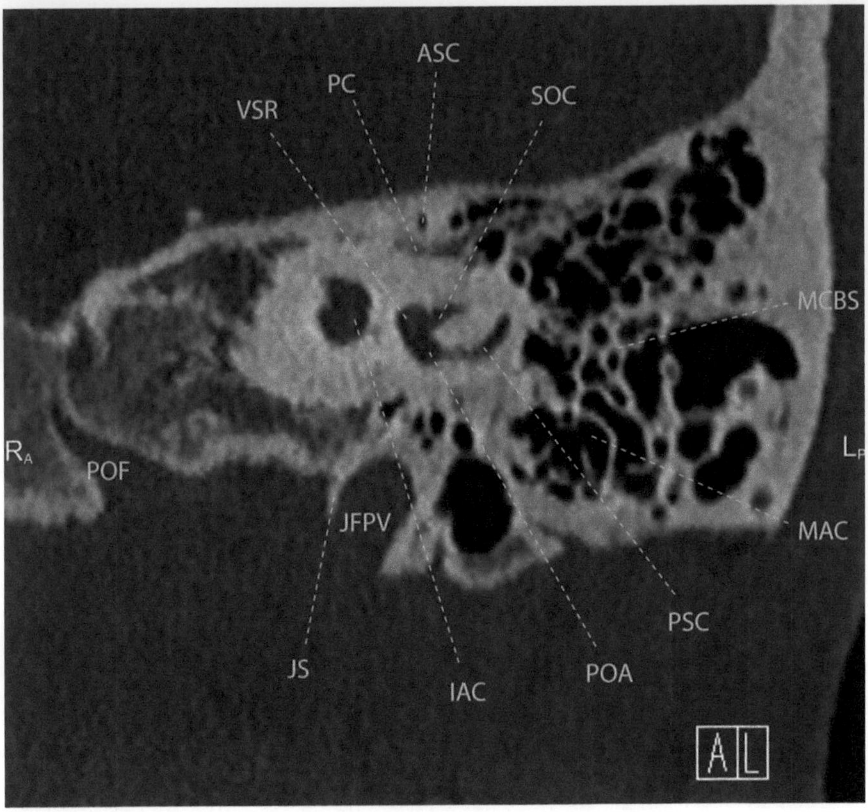

S29: HRCT Stenvers reformat

ASC	anterior semicircular canal
IAC	internal auditory canal
JFPV	jugular foramen – pars vascularis
JS	jugular spine
MAC	mastoid air cells
MCBS	mastoid cells bony septa
PC	petromastoid canal
POA	posterior osseous ampulla
POF	petrooccipital fissure
PSC	posterior semicircular canal
SOC	simple osseous crus
VSR	vestibule – spherical recess

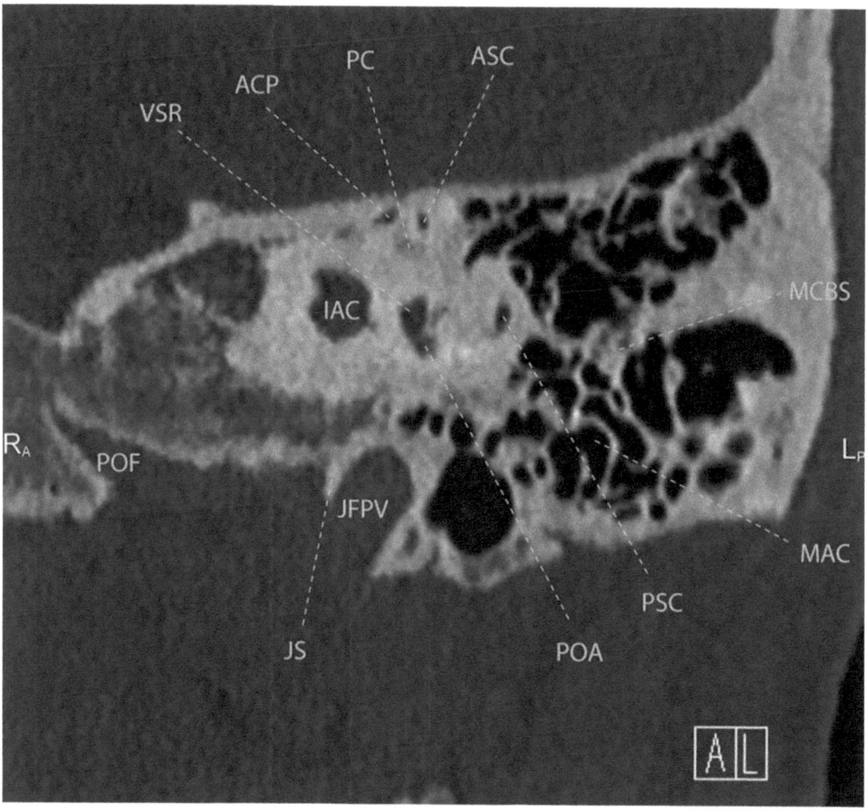

S30: HRCT Stenvers reformat

ACP	air cell of the pyramid
ASC	anterior semicircular canal
IAC	internal auditory canal
JFPV	jugular foramen – pars vascularis
JS	jugular spine
MAC	mastoid air cells
MCBS	mastoid cells bony septa
PC	petromastoid canal
POA	posterior osseous ampulla
POF	petrooccipital fissure
PSC	posterior semicircular canal
VSR	vestibule – spherical recess

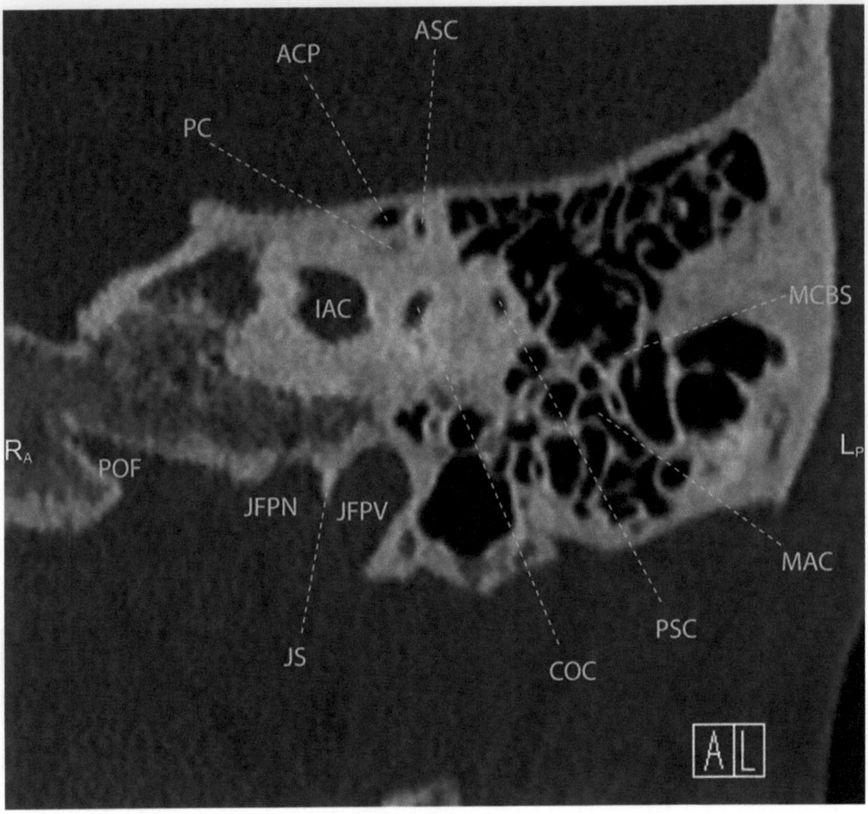

S31: HRCT Stenvers reformat

ACP	air cell of the pyramid
ASC	anterior semicircular canal
COC	common osseous crus
IAC	internal auditory canal
JFPN	jugular foramen – pars nervosa
JFPV	jugular foramen – pars vascularis
JS	jugular spine
MAC	mastoid air cells
MCBS	mastoid cells bony septa
PC	petromastoid canal
POF	petrooccipital fissure
PSC	posterior semicircular canal

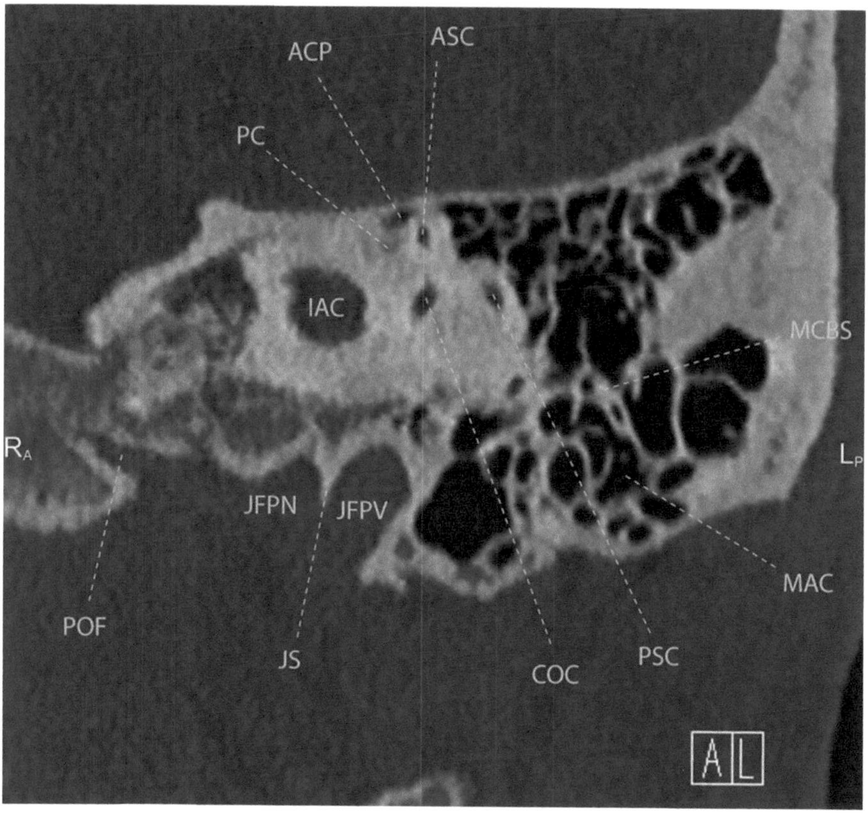

S32: HRCT Stenvers reformat

ACP	air cell of the pyramid	JS	jugular spine
ASC	anterior semicircular canal	MAC	mastoid air cells
COC	common osseous crus	MCBS	mastoid cells bony septa
IAC	internal auditory canal	PC	petromastoid canal
JFPN	jugular foramen – pars nervosa	POF	petrooccipital fissure
JFPV	jugular foramen – pars vascularis	PSC	posterior semicircular canal

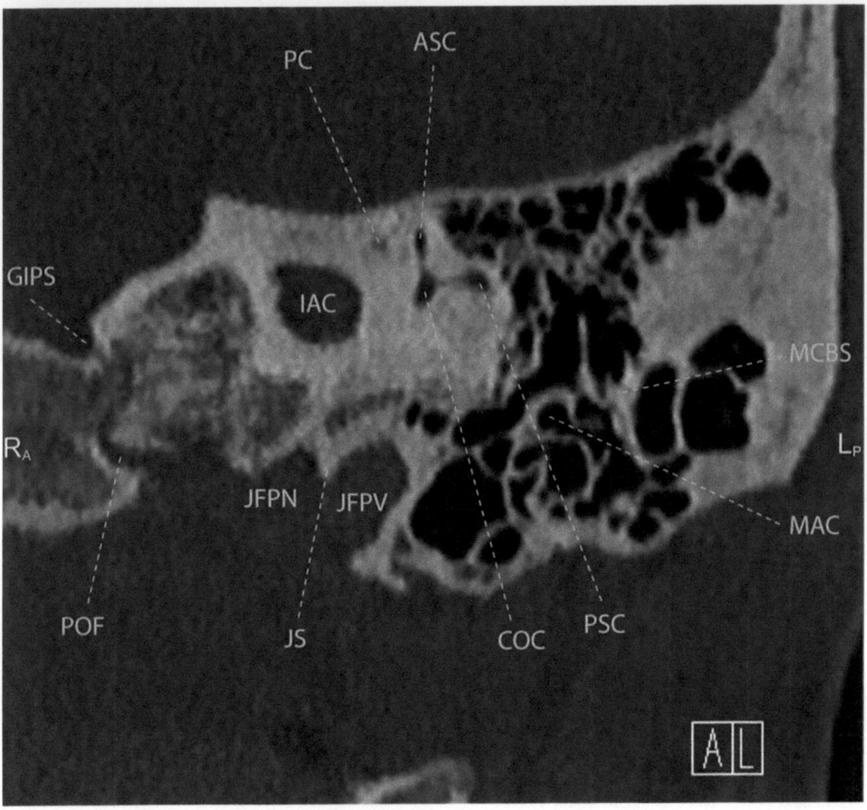

S33: HRCT Stenvers reformat

ASC	anterior semicircular canal
COC	common osseous crus
GIPS	groove of the inferior petrosal sinus
IAC	internal auditory canal
JFPN	jugular foramen – pars nervosa
JFPV	jugular foramen – pars vascularis
JS	jugular spine
MAC	mastoid air cells
MCBS	mastoid cells bony septa
PC	petromastoid canal
POF	petrooccipital fissure
PSC	posterior semicircular canal

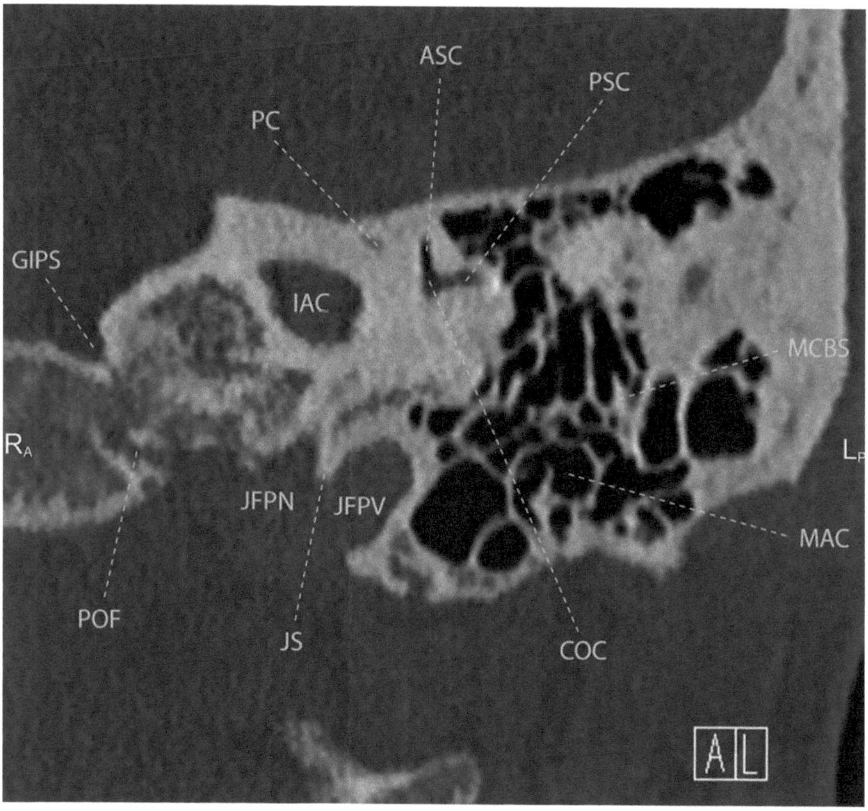

S34: HRCT Stenvers reformat

ASC	anterior semicircular canal	JS	jugular spine
COC	common osseous crus	MAC	mastoid air cells
GIPS	groove of the inferior petrosal sinus	MCBS	mastoid cells bony septa
IAC	internal auditory canal	PC	petromastoid canal
JFPN	jugular foramen – pars nervosa	POF	petrooccipital fissure
JFPV	jugular foramen – pars vascularis	PSC	posterior semicircular canal

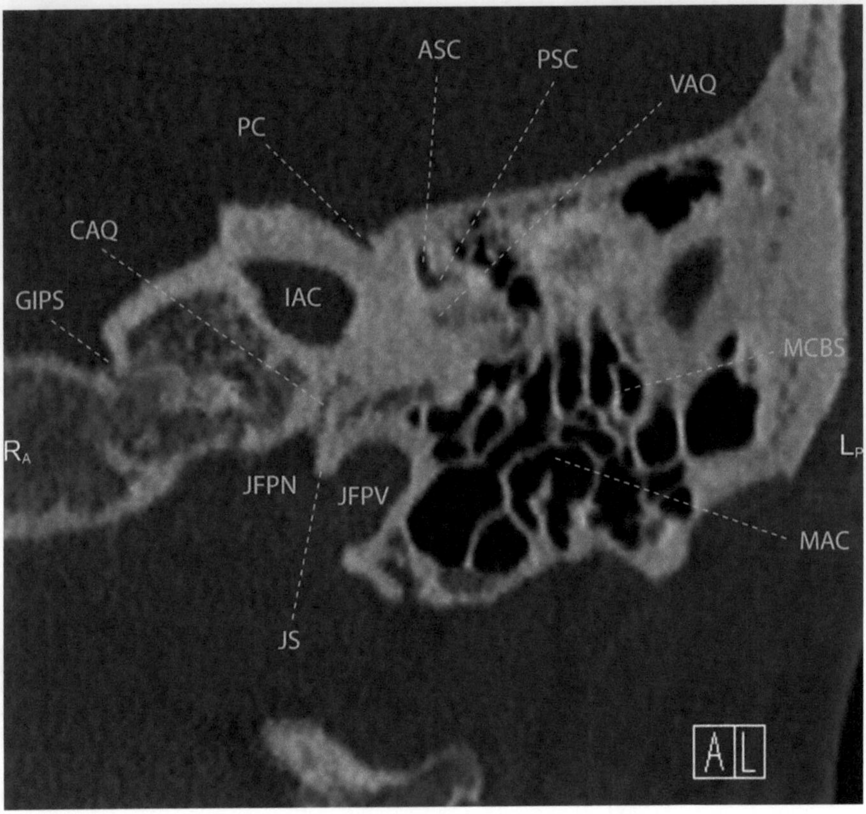

S35: HRCT Stenvers reformat

ASC	anterior semicircular canal
CAQ	cochlear aqueduct
GIPS	groove of the inferior petrosal sinus
IAC	internal auditory canal
JFPN	jugular foramen – pars nervosa
JFPV	jugular foramen – pars vascularis
JS	jugular spine
MAC	mastoid air cells
MCBS	mastoid cells bony septa
PC	petromastoid canal
PSC	posterior semicircular canal
VAQ	vestibular aqueduct

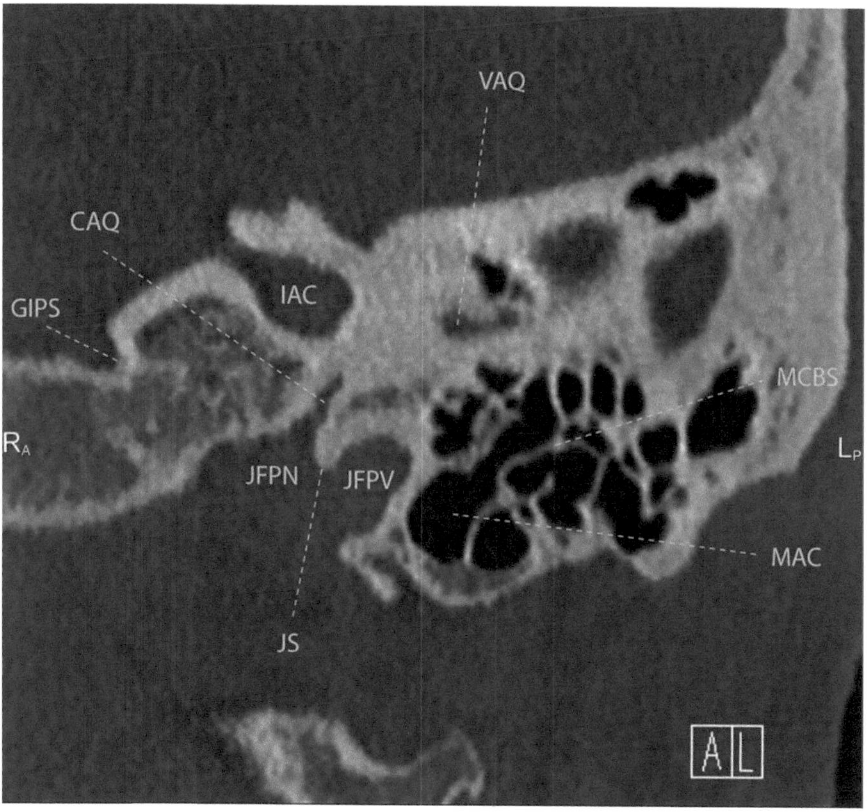

S36: HRCT Stenvers reformat

CAQ	cochlear aqueduct
GIPS	groove of the inferior petrosal sinus
IAC	internal auditory canal
JFPN	jugular foramen – pars nervosa
JFPV	jugular foramen – pars vascularis
JS	jugular spine
MAC	mastoid air cells
MCBS	mastoid cells bony septa
VAQ	vestibular aqueduct

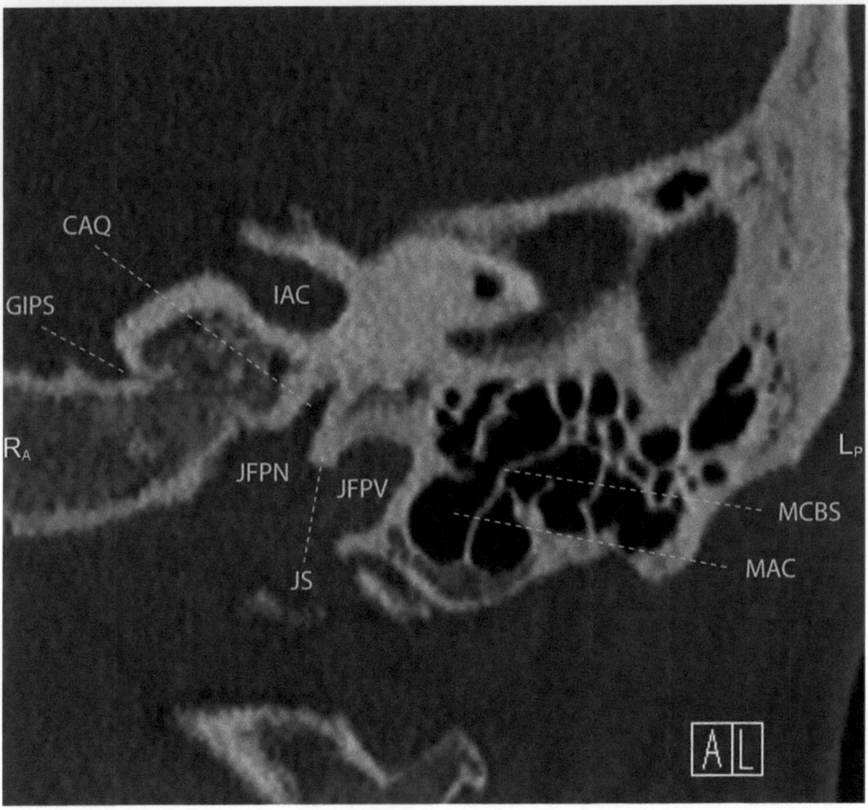

S37: HRCT Stenvers reformat

CAQ	cochlear aqueduct	JFPV	jugular foramen – pars vascularis
GIPS	groove of the inferior petrosal sinus	JS	jugular spine
IAC	internal auditory canal	MAC	mastoid air cells
JFPN	jugular foramen – pars nervosa	MCBS	mastoid cells bony septa

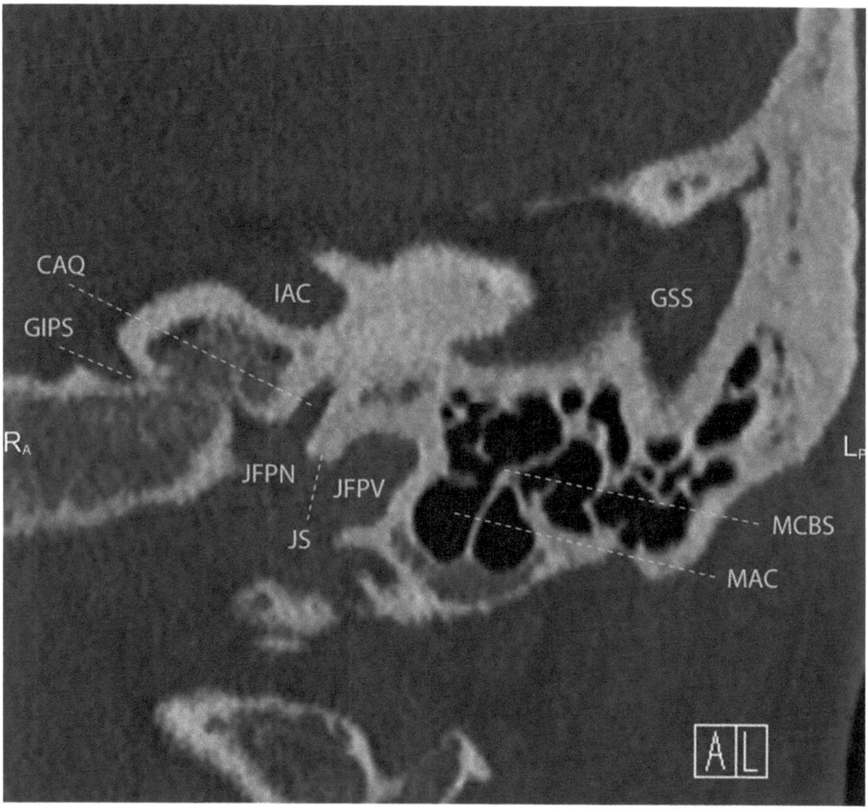

S38: HRCT Stenvers reformat

CAQ	cochlear aqueduct
GIPS	groove of the inferior petrosal sinus
GSS	groove of the sigmoid sinus
IAC	internal auditory canal
JFPN	jugular foramen – pars nervosa
JFPV	jugular foramen – pars vascularis
JS	jugular spine
MAC	mastoid air cells
MCBS	mastoid cells bony septa

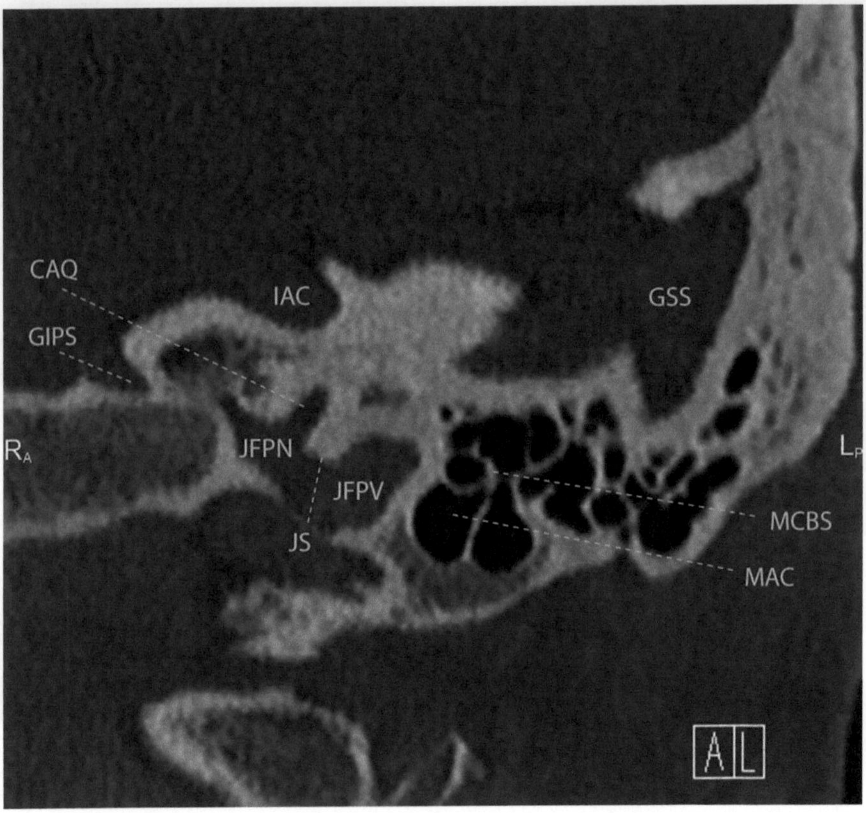

S39: HRCT Stenvers reformat

CAQ	cochlear aqueduct
GIPS	groove of the inferior petrosal sinus
GSS	groove of the sigmoid sinus
IAC	internal auditory canal
JFPN	jugular foramen – pars nervosa
JFPV	jugular foramen – pars vascularis
JS	jugular spine
MAC	mastoid air cells
MCBS	mastoid cells bony septa

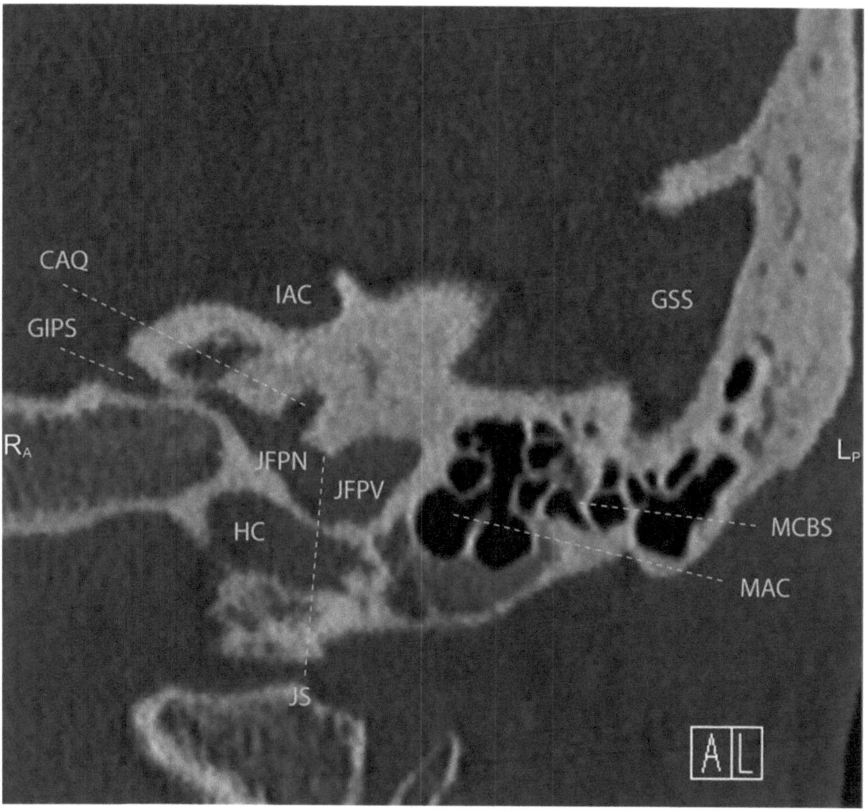

S40: HRCT Stenvers reformat

CAQ	cochlear aqueduct
GIPS	groove of the inferior petrosal sinus
GSS	groove of the sigmoid sinus
HC	hypoglossal canal
IAC	internal auditory canal
JFPN	jugular foramen – pars nervosa
JFPV	jugular foramen – pars vascularis
JS	jugular spine
MAC	mastoid air cells
MCBS	mastoid cells bony septa

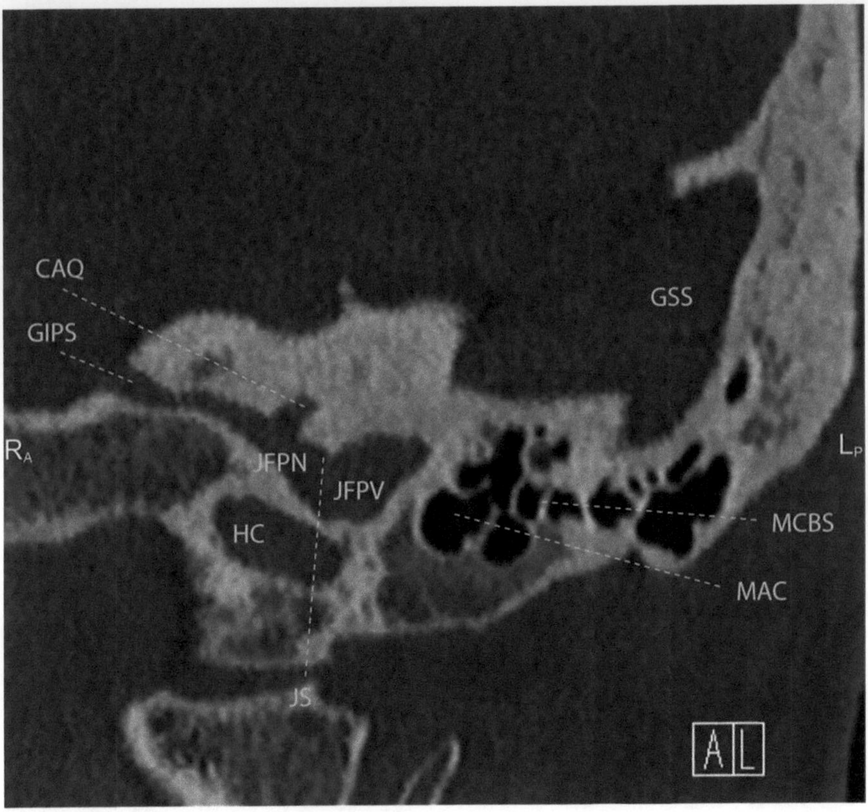

S41: HRCT Stenvers reformat

CAQ	cochlear aqueduct
GIPS	groove of the inferior petrosal sinus
GSS	groove of the sigmoid sinus
HC	hypoglossal canal
JFPN	jugular foramen – pars nervosa
JFPV	jugular foramen – pars vascularis
JS	jugular spine
MAC	mastoid air cells
MCBS	mastoid cells bony septa

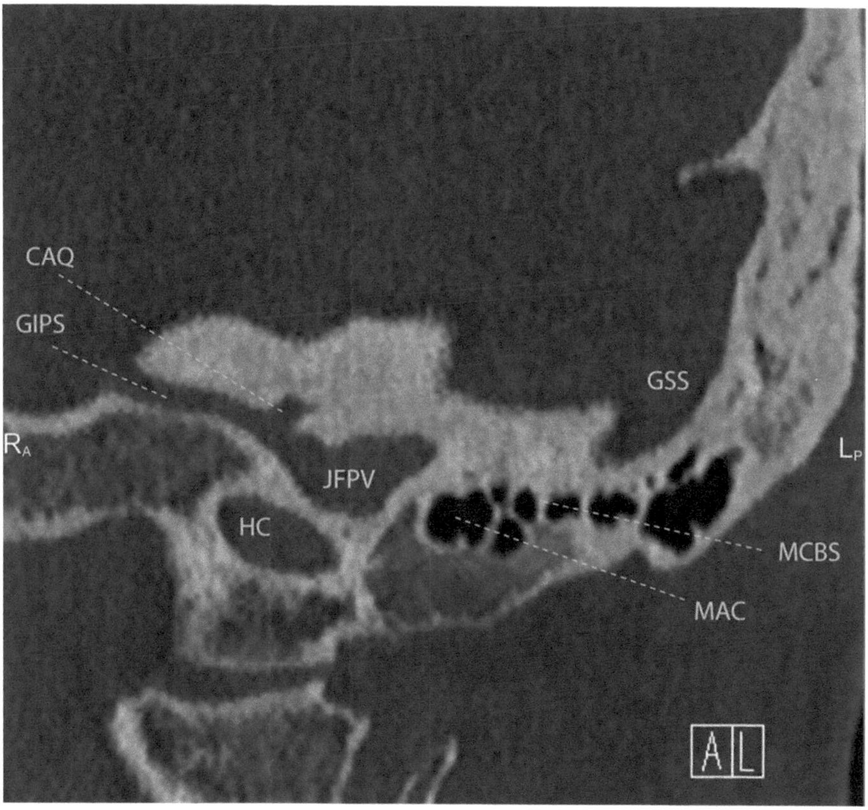

S42: HRCT Stenvers reformat

CAQ	cochlear aqueduct
GIPS	groove of the inferior petrosal sinus
GSS	groove of the sigmoid sinus
HC	hypoglossal canal
JFPV	jugular foramen – pars vascularis
MAC	mastoid air cells
MCBS	mastoid cells bony septa

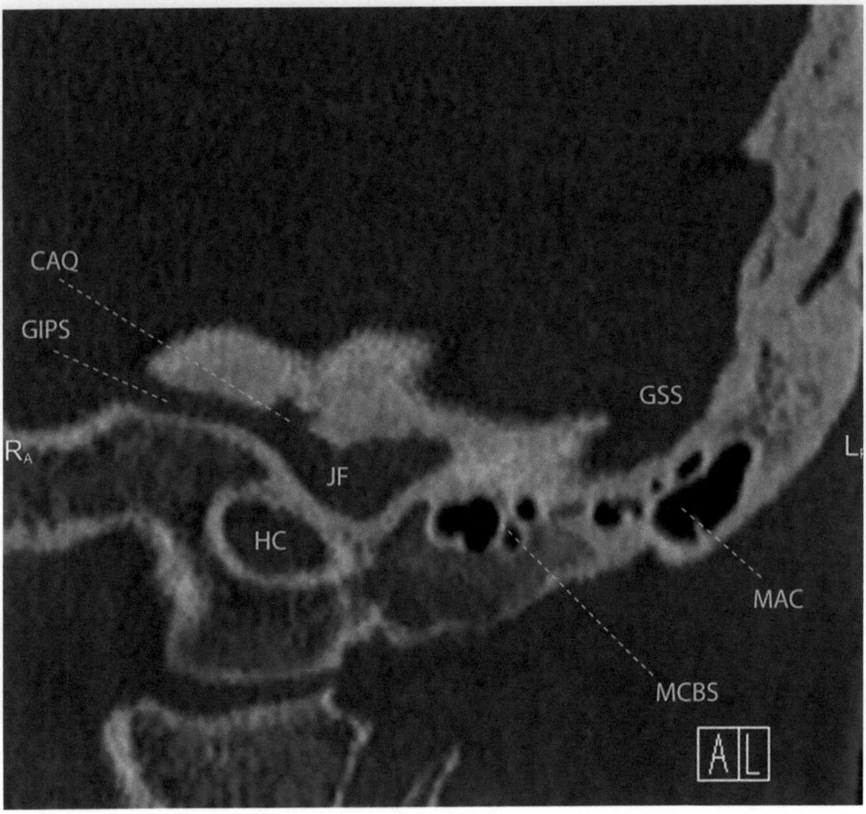

S43: HRCT Stenvers reformat

CAQ	cochlear aqueduct
GIPS	groove of the inferior petrosal sinus
GSS	groove of the sigmoid sinus
HC	hypoglossal canal
JF	jugular foramen
MAC	mastoid air cells
MCBS	mastoid cells bony septa

Chapter 5
MRI Axial Sections

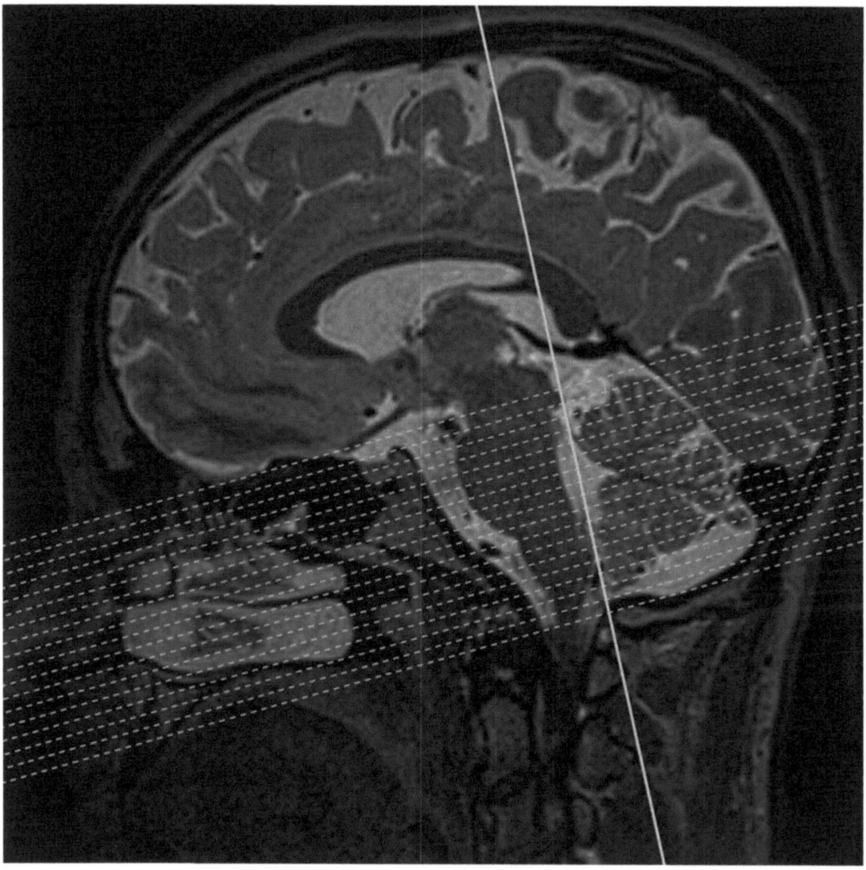

M00: MRI axial sections

Standard axial MRI section is perpendicular to the posterior margin of the brainstem defined on a midsagittal image.

J. Kopřiva, J. Žižka, *Temporal Bone CT and MRI Anatomy: A Guide to 3D Volumetric Acquisitions*, DOI 10.1007/978-3-319-08242-4_5

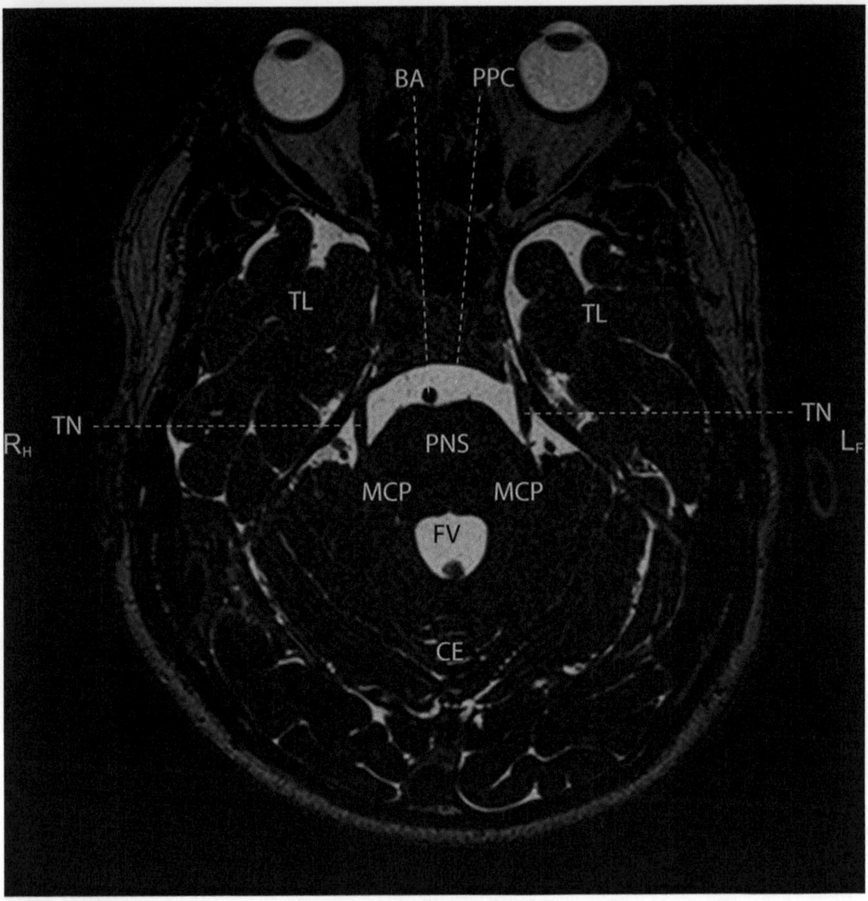

M01: MRI axial sections

BA	basilar artery	PNS	pons
CE	cerebellum	PPC	prepontine cistern
FV	fourth ventricle	TL	temporal lobe
MCP	middle cerebellar peduncle	TN	trigeminal nerve

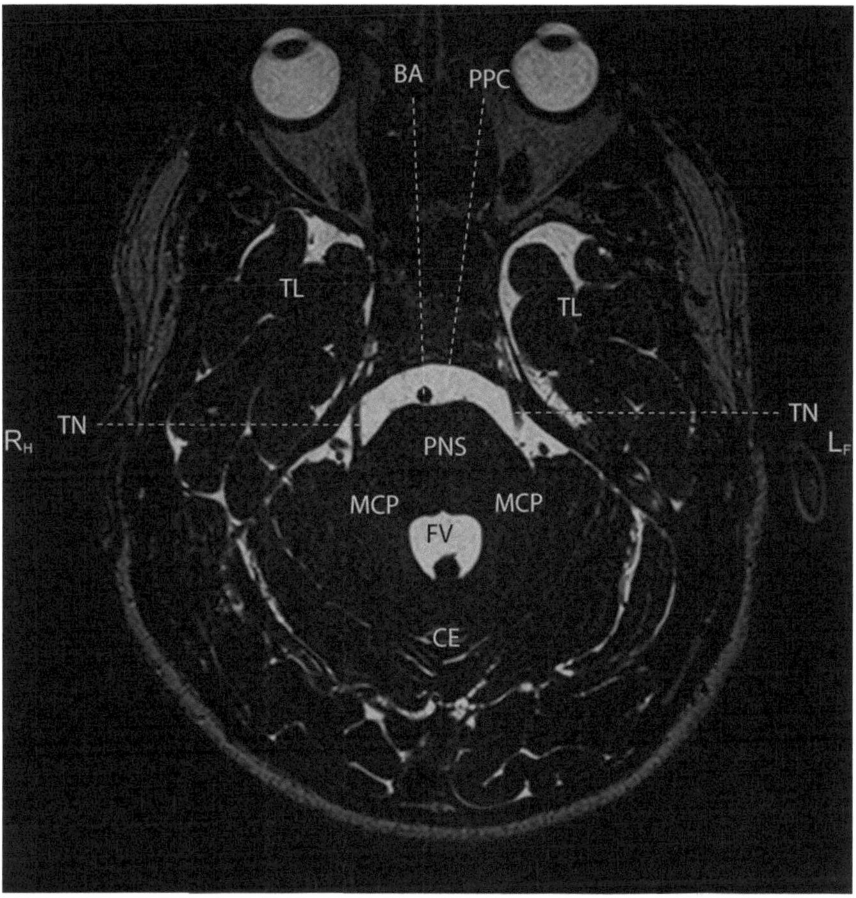

M02: MRI axial sections

BA	basilar artery	PNS	pons
CE	cerebellum	PPC	prepontine cistern
FV	fourth ventricle	TL	temporal lobe
MCP	middle cerebellar peduncle	TN	trigeminal nerve

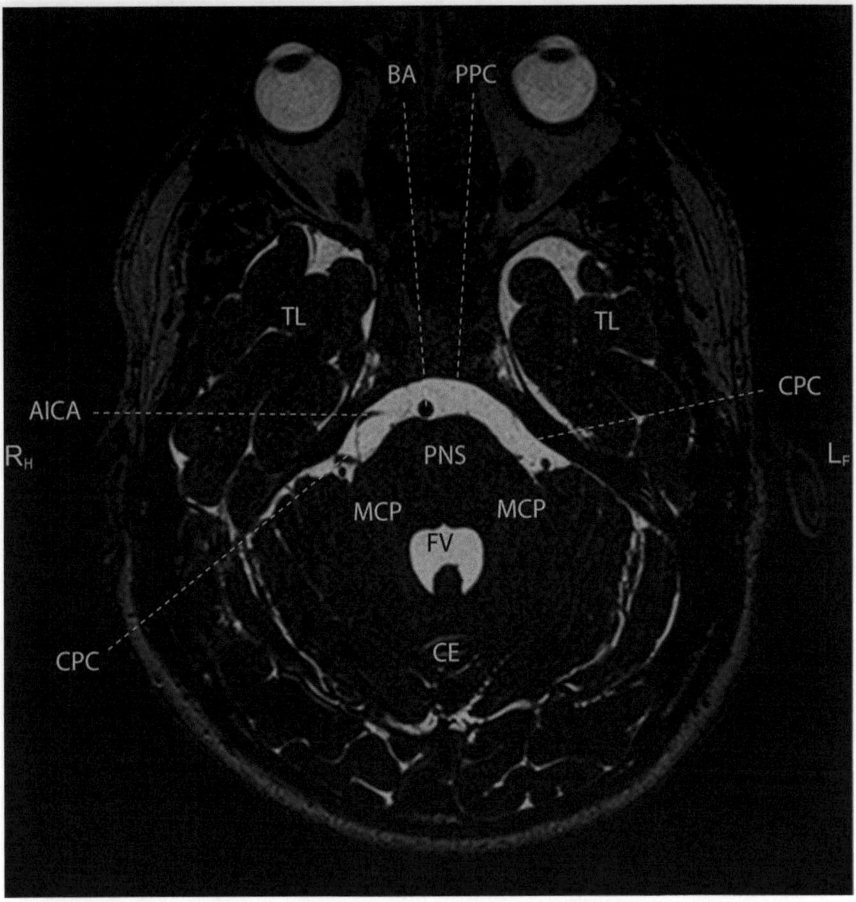

M03: MRI axial sections

AICA	anterior inferior cerebellar artery	MCP	middle cerebellar peduncle
BA	basilar artery	PNS	pons
CE	cerebellum	PPC	prepontine cistern
CPC	cerebellopontine cistern	TL	temporal lobe
FV	fourth ventricle		

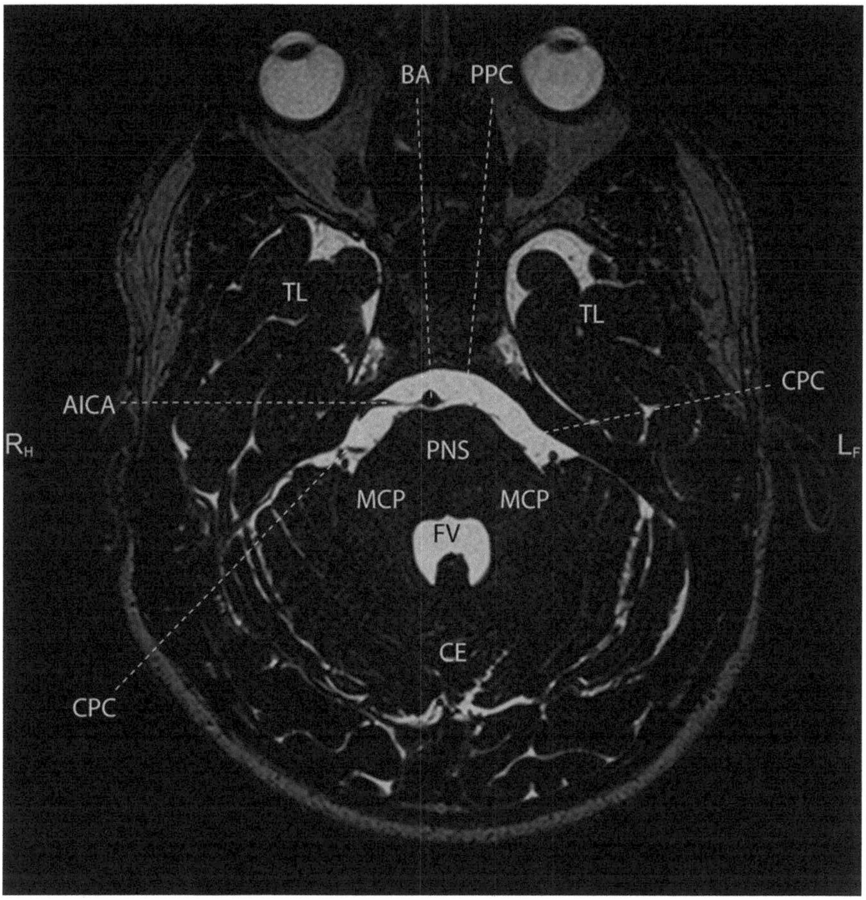

M04: MRI axial sections

AICA	anterior inferior cerebellar artery	MCP	middle cerebellar peduncle
BA	basilar artery	PNS	pons
CE	cerebellum	PPC	prepontine cistern
CPC	cerebellopontine cistern	TL	temporal lobe
FV	fourth ventricle		

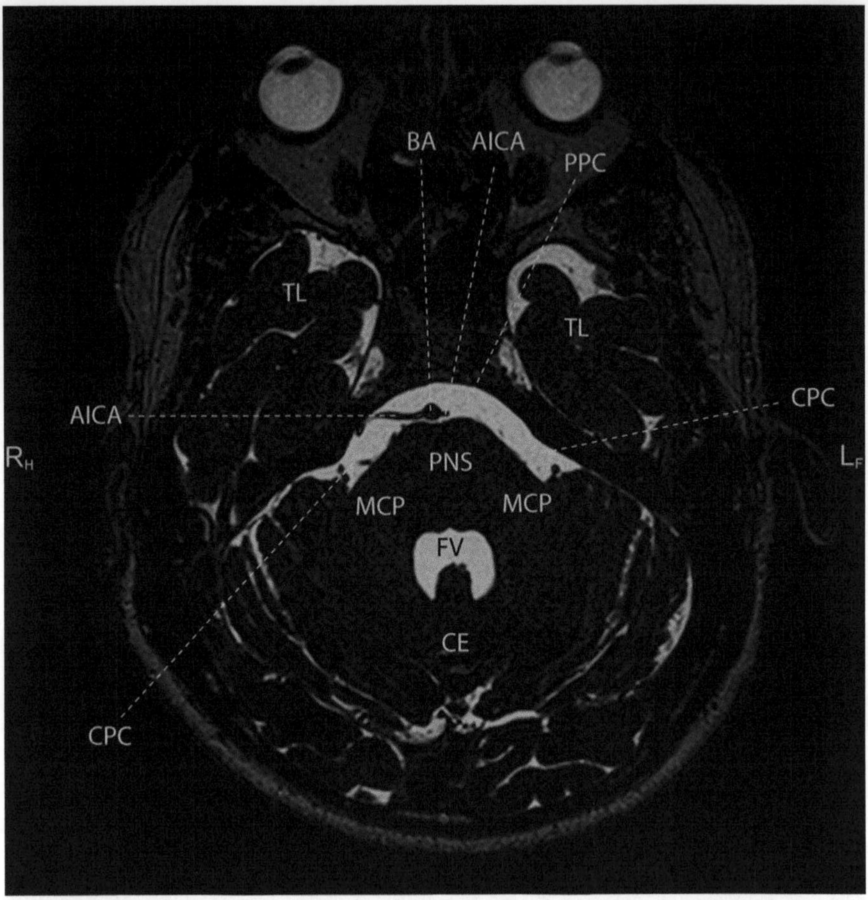

M05: MRI axial sections

AICA	anterior inferior cerebellar artery	MCP	middle cerebellar peduncle
BA	basilar artery	PNS	pons
CE	cerebellum	PPC	prepontine cistern
CPC	cerebellopontine cistern	TL	temporal lobe
FV	fourth ventricle		

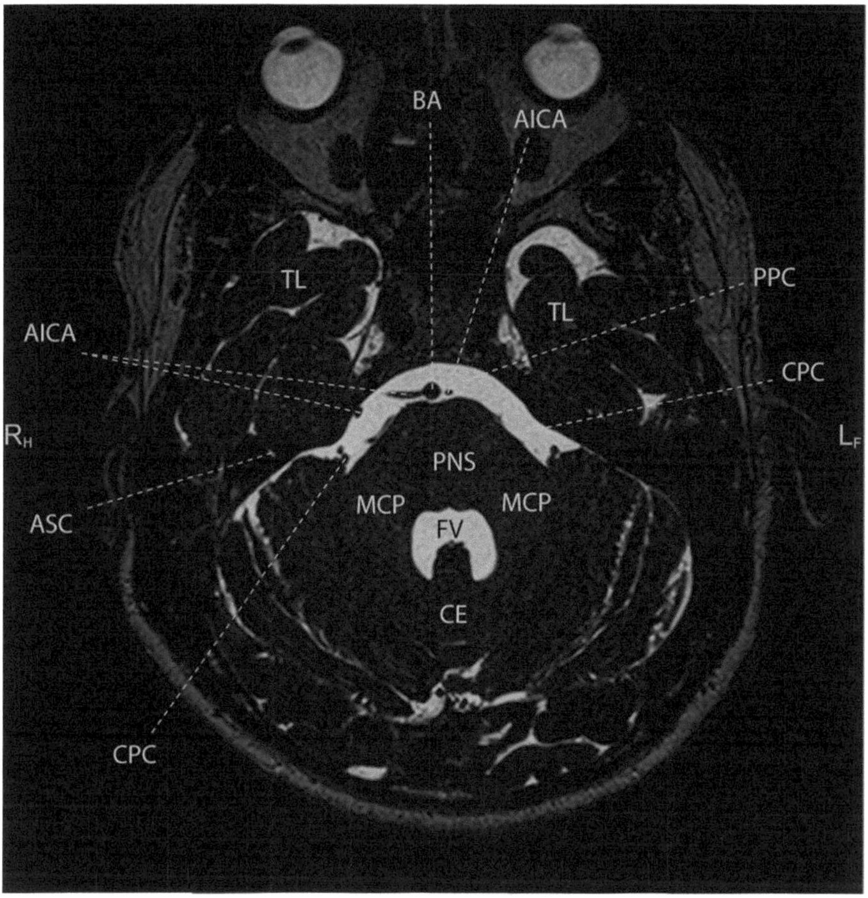

M06: MRI axial sections

AICA	anterior inferior cerebellar artery	FV	fourth ventricle
ASC	anterior semicircular canal	MCP	middle cerebellar peduncle
BA	basilar artery	PNS	pons
CE	cerebellum	PPC	prepontine cistern
CPC	cerebellopontine cistern	TL	temporal lobe

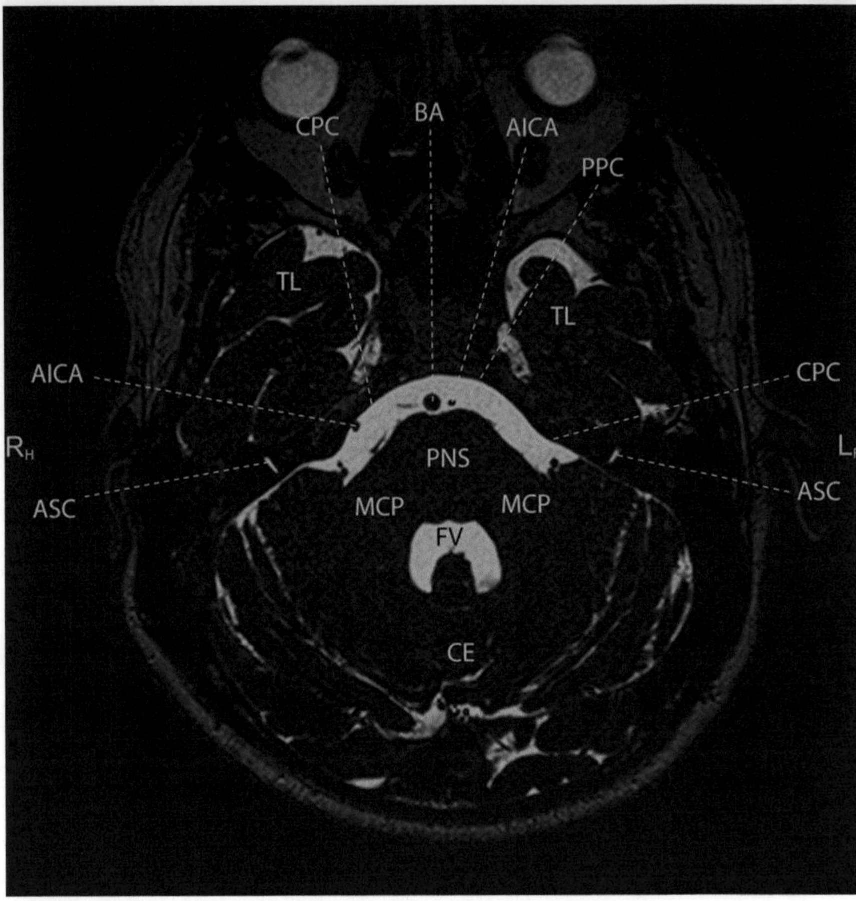

M07: MRI axial sections

AICA	anterior inferior cerebellar artery	FV	fourth ventricle
ASC	anterior semicircular canal	MCP	middle cerebellar peduncle
BA	basilar artery	PNS	pons
CE	cerebellum	PPC	prepontine cistern
CPC	cerebellopontine cistern	TL	temporal lobe

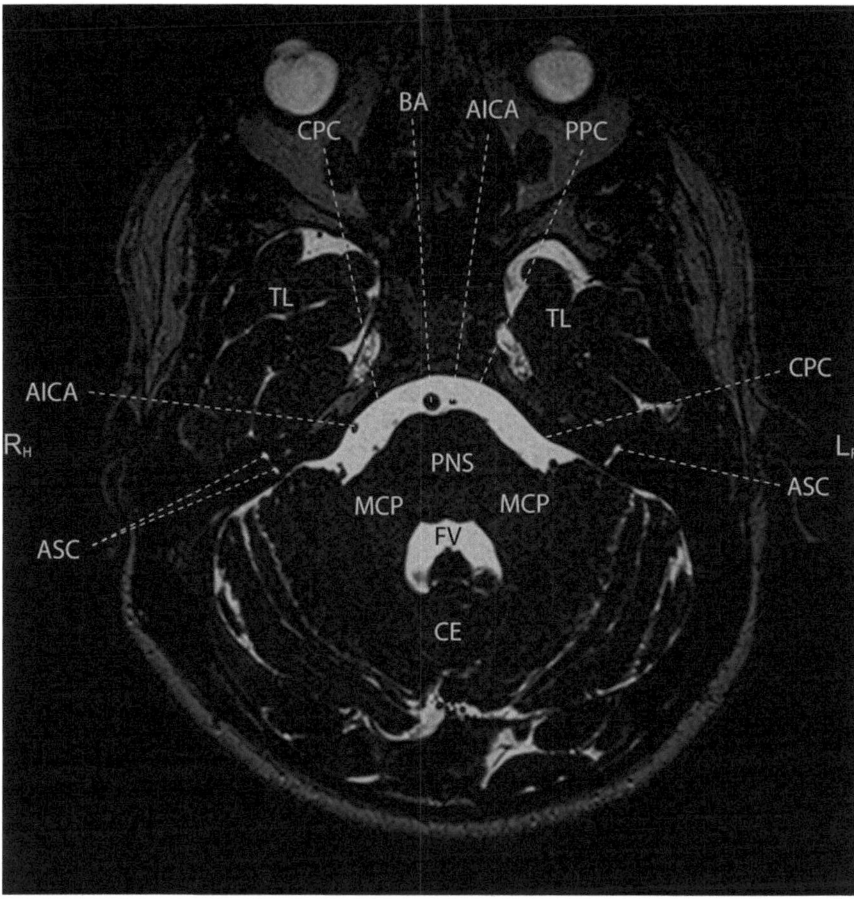

M08: MRI axial sections

AICA	anterior inferior cerebellar artery	FV	fourth ventricle
ASC	anterior semicircular canal	MCP	middle cerebellar peduncle
BA	basilar artery	PNS	pons
CE	cerebellum	PPC	prepontine cistern
CPC	cerebellopontine cistern	TL	temporal lobe

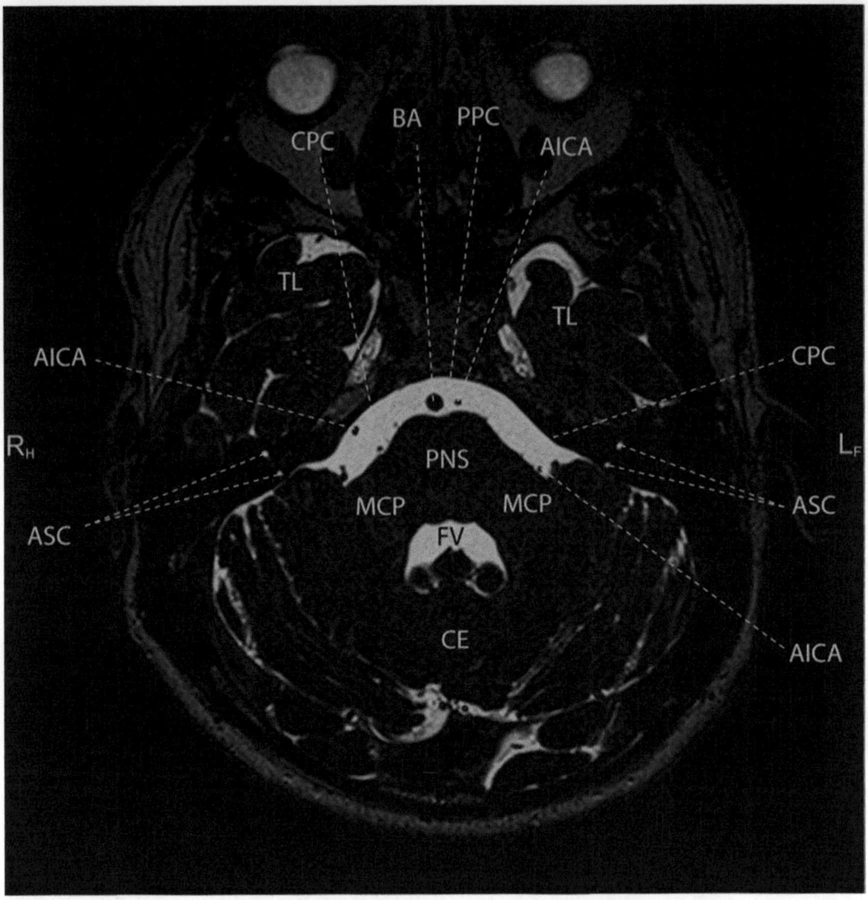

M09: MRI axial sections

AICA	anterior inferior cerebellar artery	FV	fourth ventricle
ASC	anterior semicircular canal	MCP	middle cerebellar peduncle
BA	basilar artery	PNS	pons
CE	cerebellum	PPC	prepontine cistern
CPC	cerebellopontine cistern	TL	temporal lobe

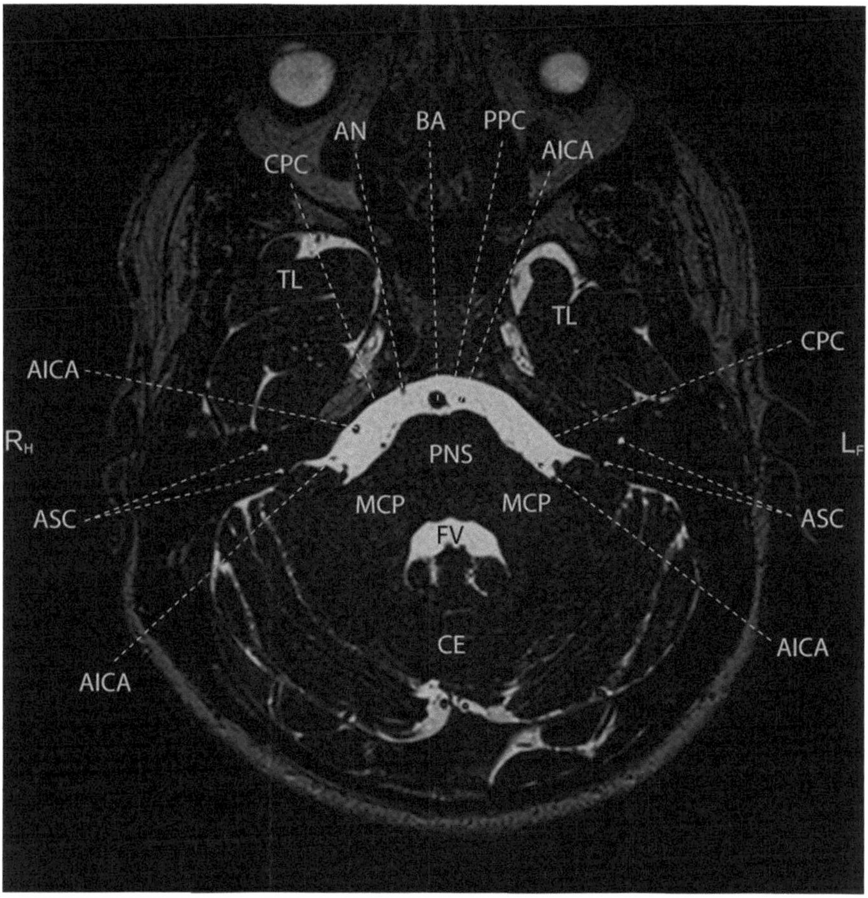

M10: MRI axial sections

AICA	anterior inferior cerebellar artery	FV	fourth ventricle
AN	abducens nerve	MCP	middle cerebellar peduncle
ASC	anterior semicircular canal	PNS	pons
BA	basilar artery	PPC	prepontine cistern
CE	cerebellum	TL	temporal lobe
CPC	cerebellopontine cistern		

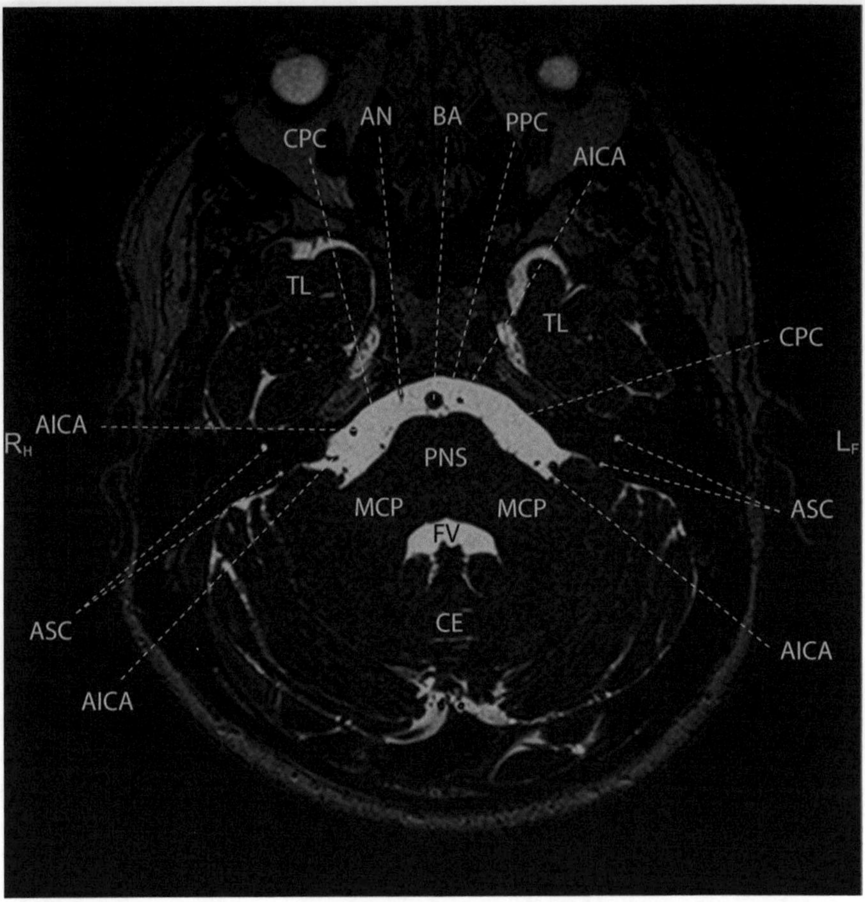

M11: MRI axial sections

AICA	anterior inferior cerebellar artery	FV	fourth ventricle
AN	abducens nerve	MCP	middle cerebellar peduncle
ASC	anterior semicircular canal	PNS	pons
BA	basilar artery	PPC	prepontine cistern
CE	cerebellum	TL	temporal lobe
CPC	cerebellopontine cistern		

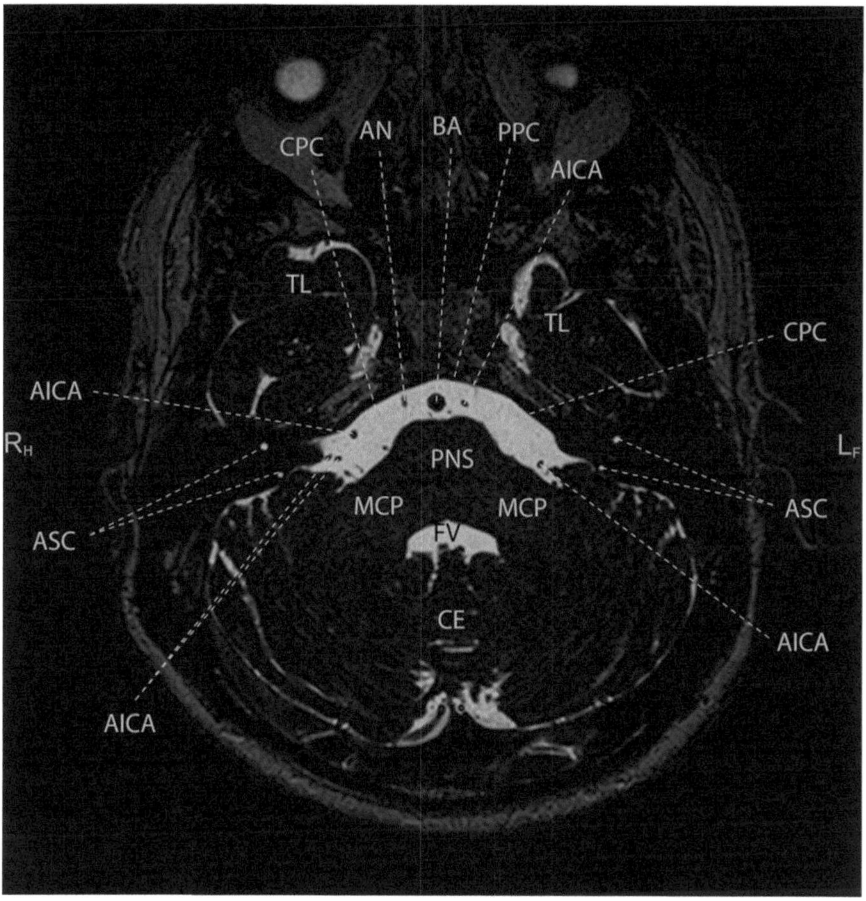

M12: MRI axial sections

AICA	anterior inferior cerebellar artery	FV	fourth ventricle
AN	abducens nerve	MCP	middle cerebellar peduncle
ASC	anterior semicircular canal	PNS	pons
BA	basilar artery	PPC	prepontine cistern
CE	cerebellum	TL	temporal lobe
CPC	cerebellopontine cistern		

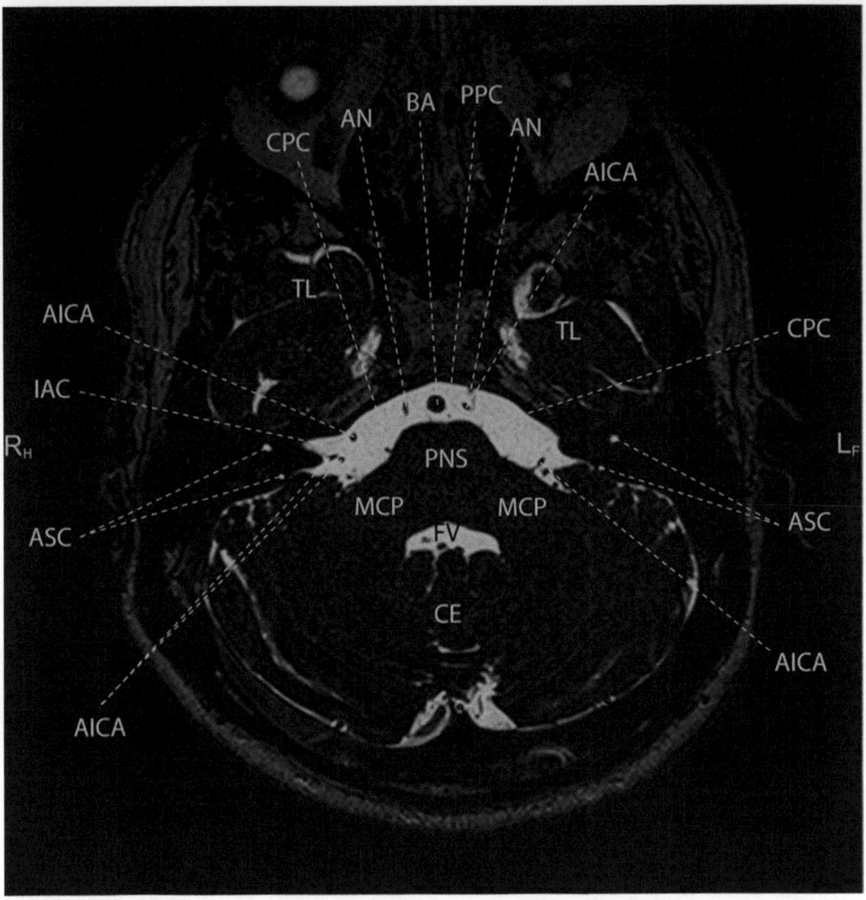

M13: MRI axial sections

AICA	anterior inferior cerebellar artery	FV	fourth ventricle
AN	abducens nerve	IAC	internal auditory canal
ASC	anterior semicircular canal	MCP	middle cerebellar peduncle
BA	basilar artery	PNS	pons
CE	cerebellum	PPC	prepontine cistern
CPC	cerebellopontine cistern	TL	temporal lobe

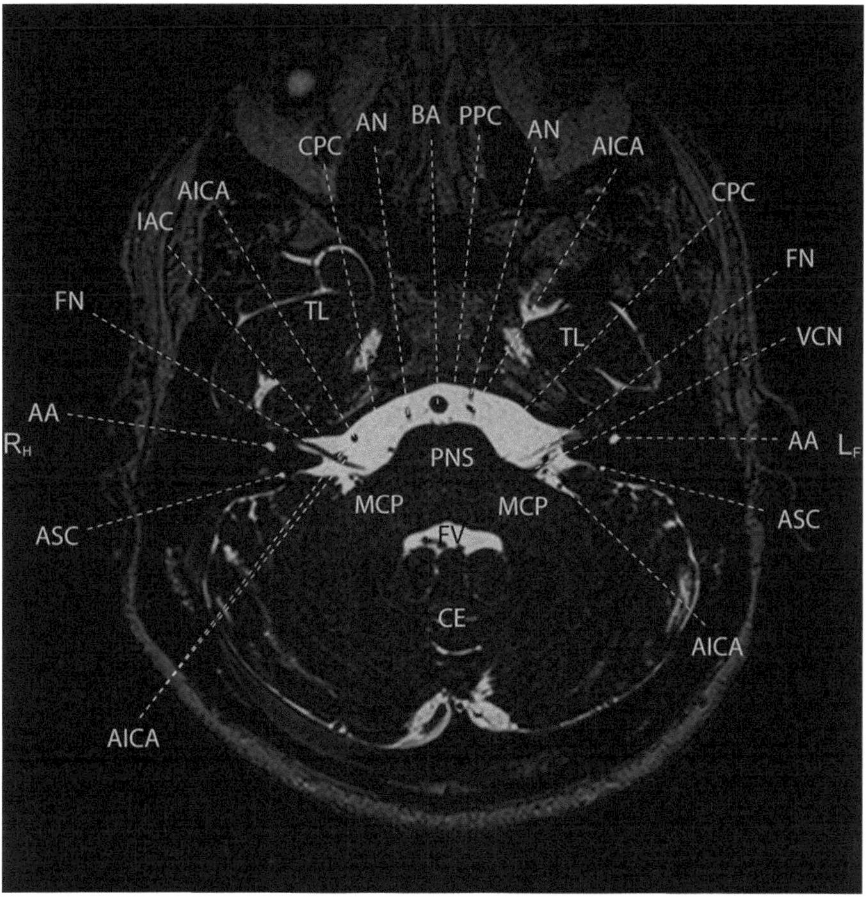

M14: MRI axial sections

AA	anterior ampulla	FV	fourth ventricle
AICA	anterior inferior cerebellar artery	IAC	internal auditory canal
AN	abducens nerve	MCP	middle cerebellar peduncle
ASC	anterior semicircular canal	PNS	pons
BA	basilar artery	PPC	prepontine cistern
CE	cerebellum	TL	temporal lobe
CPC	cerebellopontine cistern	VCN	vestibulocochlear nerve
FN	facial nerve		

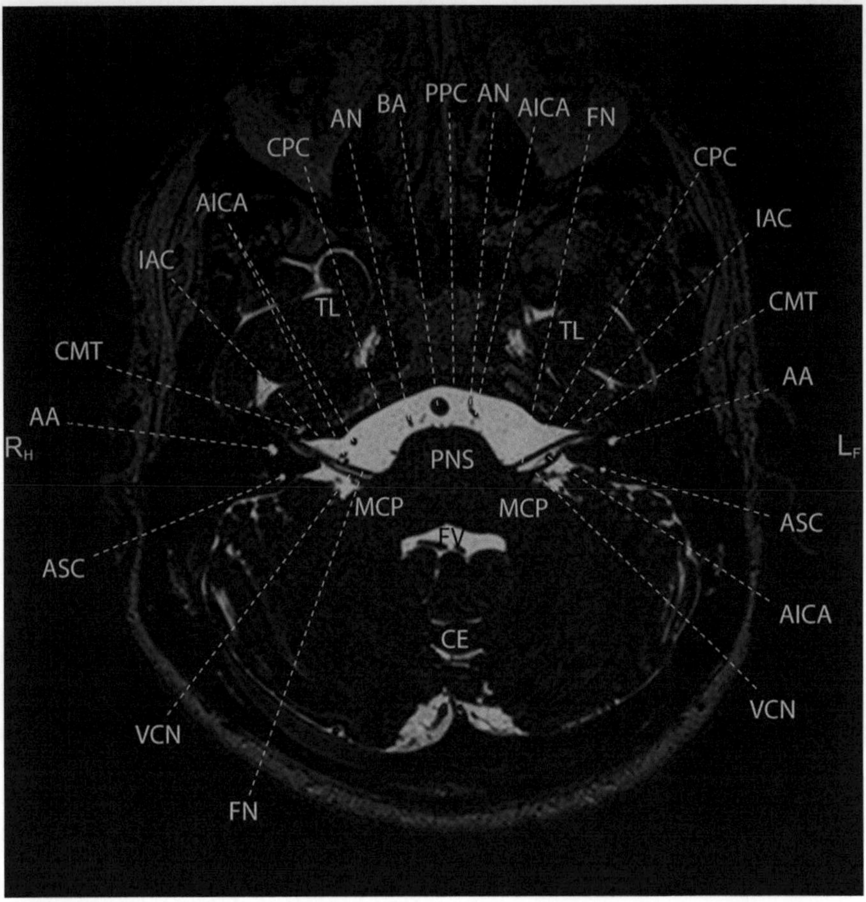

M15: MRI axial sections

AA	anterior ampulla	FN	facial nerve
AICA	anterior inferior cerebellar artery	FV	fourth ventricle
AN	abducens nerve	IAC	internal auditory canal
ASC	anterior semicircular canal	MCP	middle cerebellar peduncle
BA	basilar artery	PNS	pons
CE	cerebellum	PPC	prepontine cistern
CMT	cochlea – middle turn	TL	temporal lobe
CPC	cerebellopontine cistern	VCN	vestibulocochlear nerve

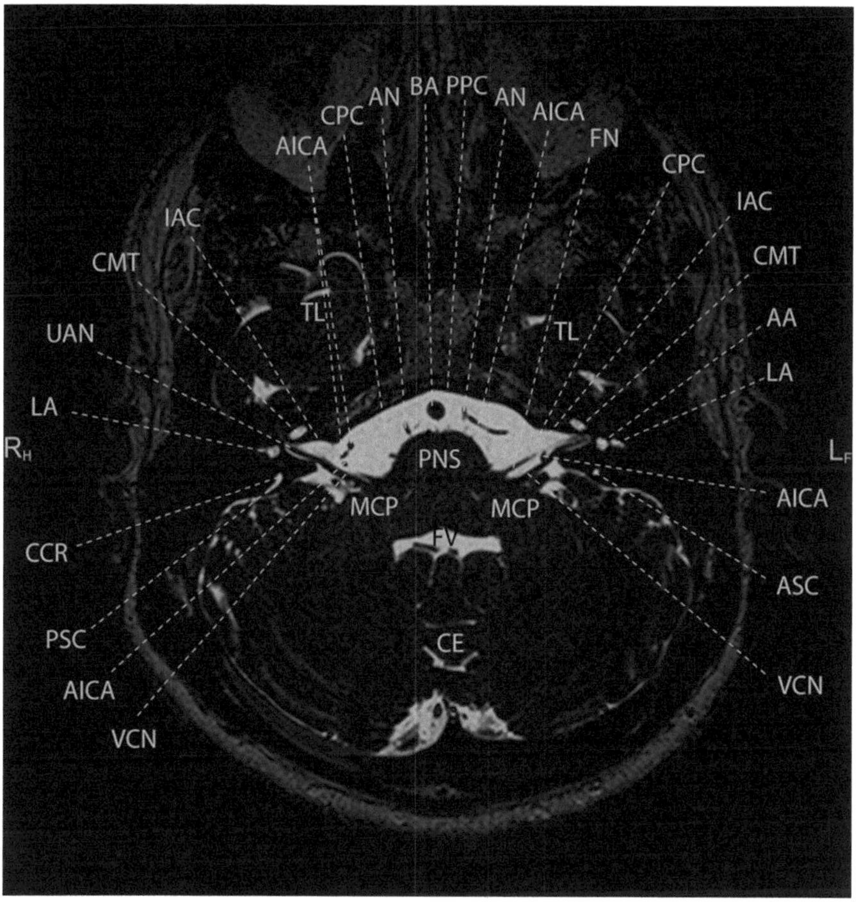

M16: MRI axial sections

AA	anterior ampulla	FV	fourth ventricle
AICA	anterior inferior cerebellar artery	IAC	internal auditory canal
AN	abducens nerve	LA	lateral ampulla
ASC	anterior semicircular canal	MCP	middle cerebellar peduncle
BA	basilar artery	PNS	pons
CCR	common crus	PPC	prepontine cistern
CE	cerebellum	PSC	posterior semicircular canal
CMT	cochlea – middle turn	TL	temporal lobe
CPC	cerebellopontine cistern	UAN	utriculo-ampullar nerve
FN	facial nerve	VCN	vestibulocochlear nerve

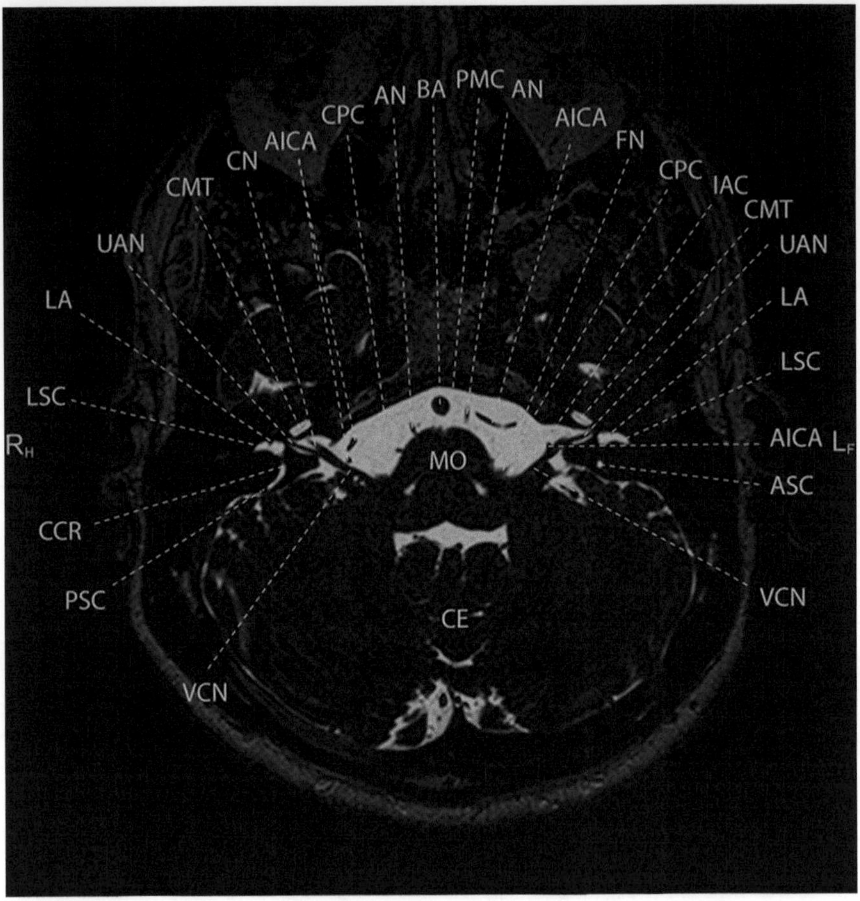

M17: MRI axial sections

AICA	anterior inferior cerebellar artery	FN	facial nerve
AN	abducens nerve	IAC	internal auditory canal
ASC	anterior semicircular canal	LA	lateral ampulla
BA	basilar artery	LSC	lateral semicircular canal
CCR	common crus	MO	medulla oblongata
CE	cerebellum	PMC	premedullary cistern
CMT	cochlea – middle turn	PSC	posterior semicircular canal
CN	cochlear nerve	UAN	utriculo-ampullar nerve
CPC	cerebellopontine cistern	VCN	vestibulocochlear nerve

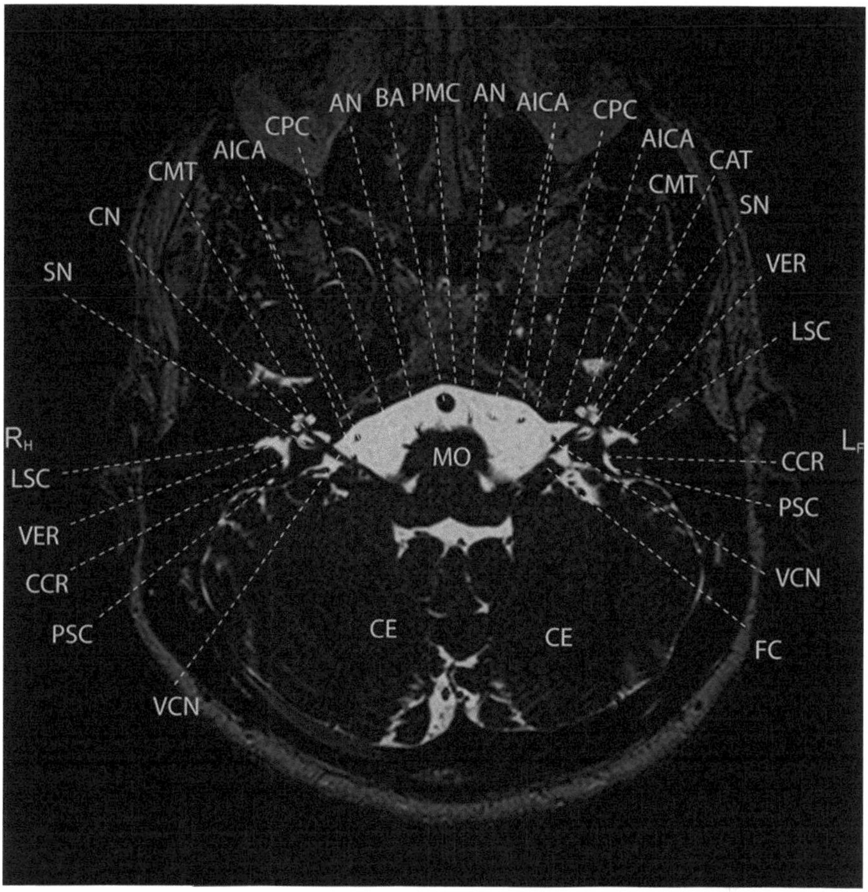

M18: MRI axial sections

AICA	anterior inferior cerebellar artery	FC	flocculus
AN	abducens nerve	LSC	lateral semicircular canal
BA	basilar artery	MO	medulla oblongata
CAT	cochlea – apical turn	PMC	premedullary cistern
CCR	common crus	PSC	posterior semicircular canal
CE	cerebellum	SN	saccular nerve
CMT	cochlea – middle turn	VCN	vestibulocochlear nerve
CN	cochlear nerve	VER	vestibule – elliptical recess
CPC	cerebellopontine cistern		

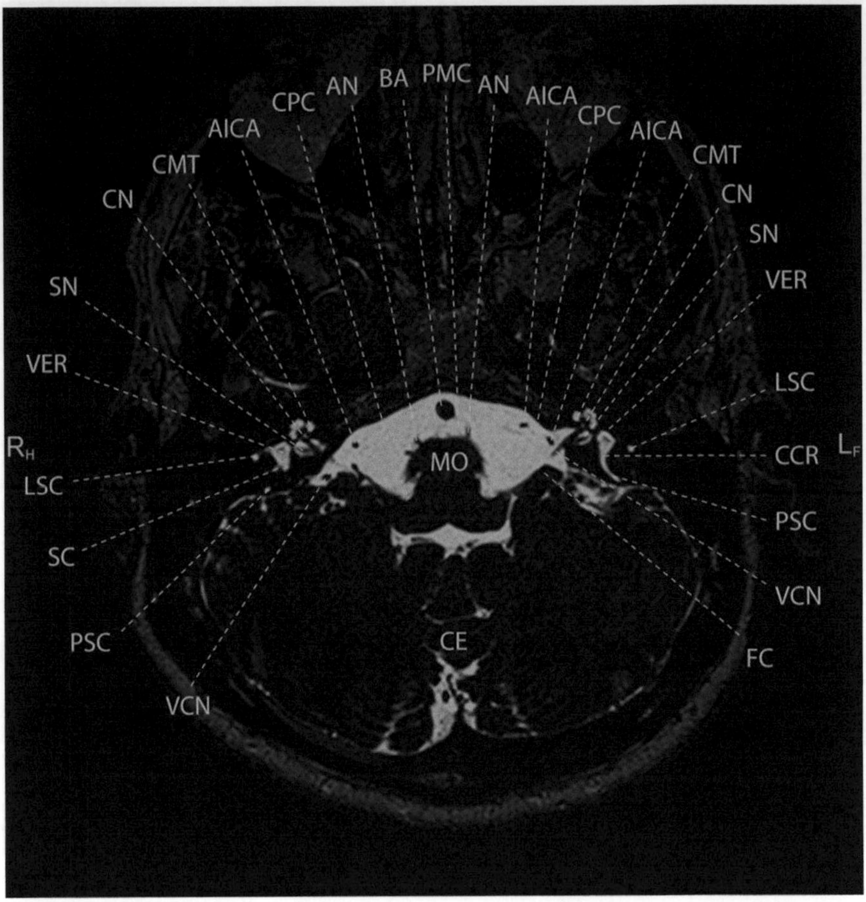

M19: MRI axial sections

AICA	anterior inferior cerebellar artery	LSC	lateral semicircular canal
AN	abducens nerve	MO	medulla oblongata
BA	basilar artery	PMC	premedullary cistern
CCR	common crus	PSC	posterior semicircular canal
CE	cerebellum	SC	simple crus
CMT	cochlea – middle turn	SN	saccular nerve
CN	cochlear nerve	VCN	vestibulocochlear nerve
CPC	cerebellopontine cistern	VER	vestibule – elliptical recess
FC	flocculus		

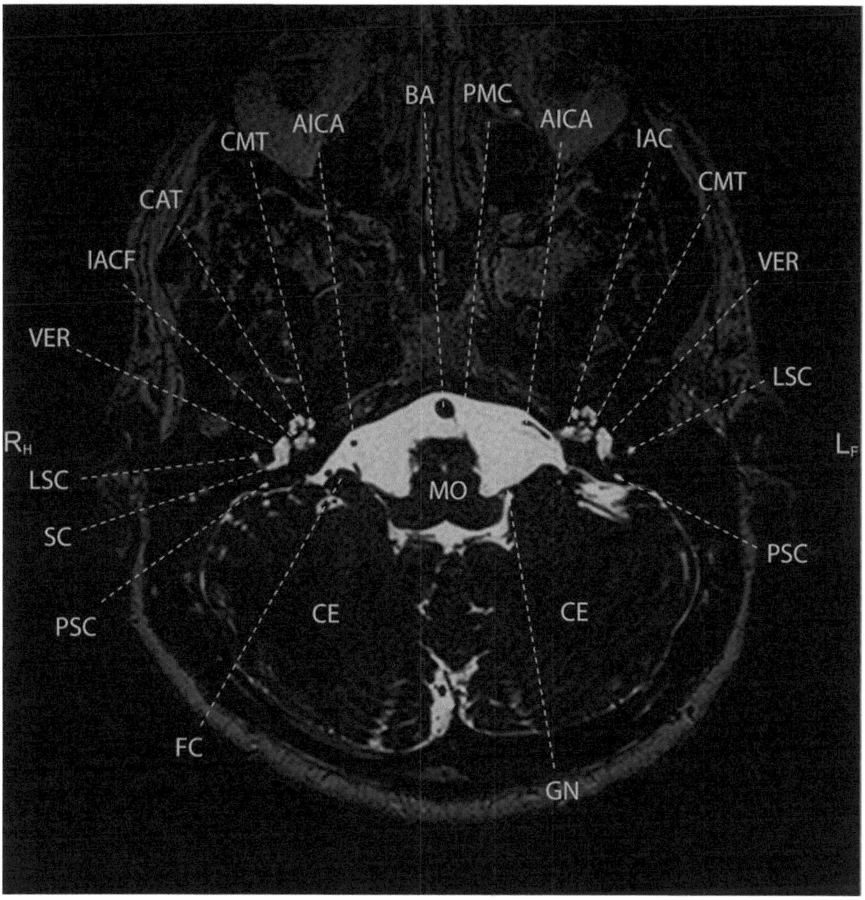

M20: MRI axial sections

AICA	anterior inferior cerebellar artery	IACF	internal auditory canal – fundus
BA	basilar artery	LSC	lateral semicircular canal
CAT	cochlea – apical turn	MO	medulla oblongata
CE	cerebellum	PMC	premedullary cistern
CMT	cochlea – middle turn	PSC	posterior semicircular canal
FC	flocculus	SC	simple crus
GN	glossopharyngeal nerve	VER	vestibule – elliptical recess
IAC	internal auditory canal		

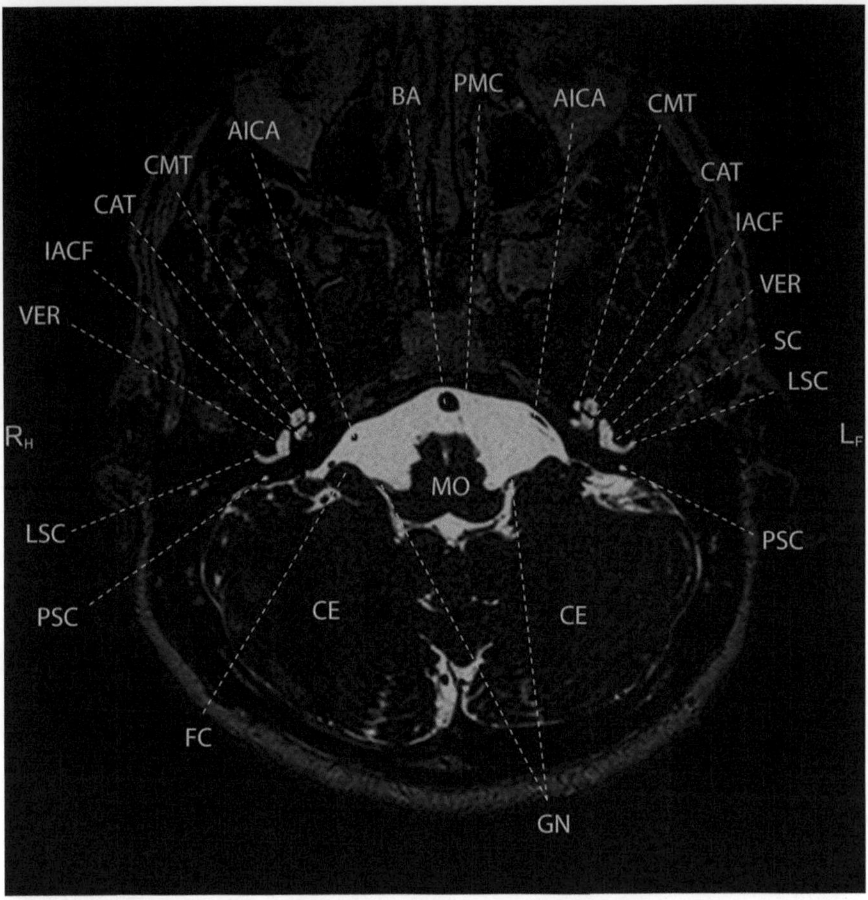

M21: MRI axial sections

AICA	anterior inferior cerebellar artery	IACF	internal auditory canal – fundus
BA	basilar artery	LSC	lateral semicircular canal
CAT	cochlea – apical turn	MO	medulla oblongata
CE	cerebellum	PMC	premedullary cistern
CMT	cochlea – middle turn	PSC	posterior semicircular canal
FC	flocculus	SC	simple crus
GN	glossopharyngeal nerve	VER	vestibule – elliptical recess

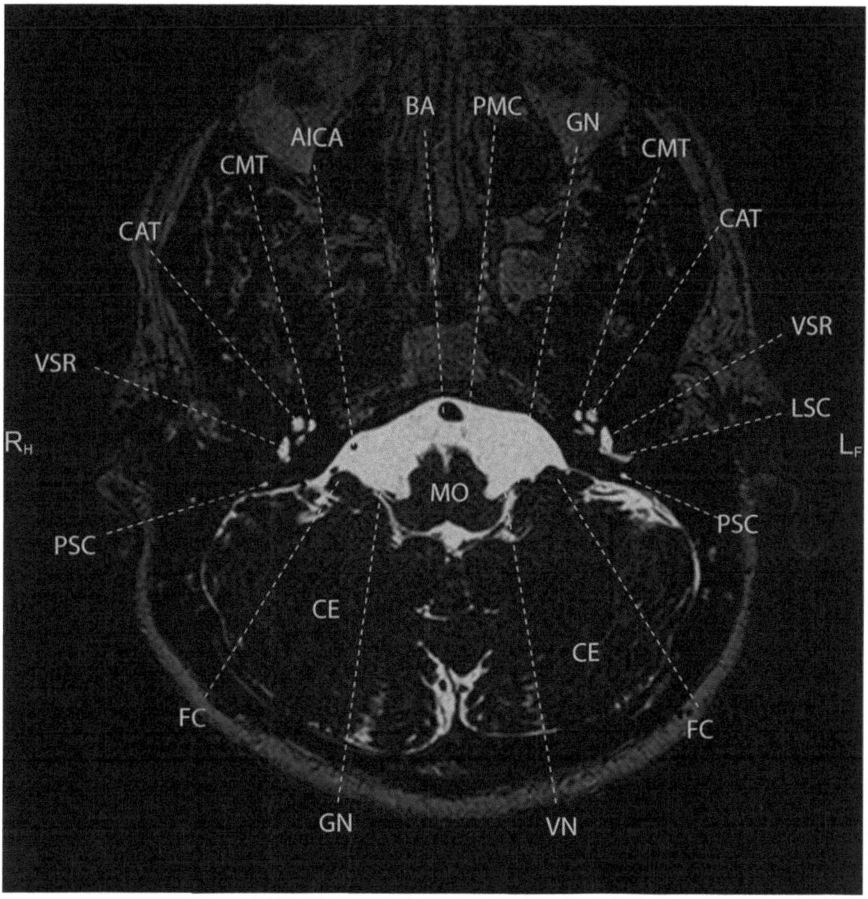

M22: MRI axial sections

AICA	anterior inferior cerebellar artery	LSC	lateral semicircular canal
BA	basilar artery	MO	medulla oblongata
CAT	cochlea – apical turn	PMC	premedullary cistern
CE	cerebellum	PSC	posterior semicircular canal
CMT	cochlea – middle turn	VN	vagus nerve
FC	flocculus	VSR	vestibule – spherical recess
GN	glossopharyngeal nerve		

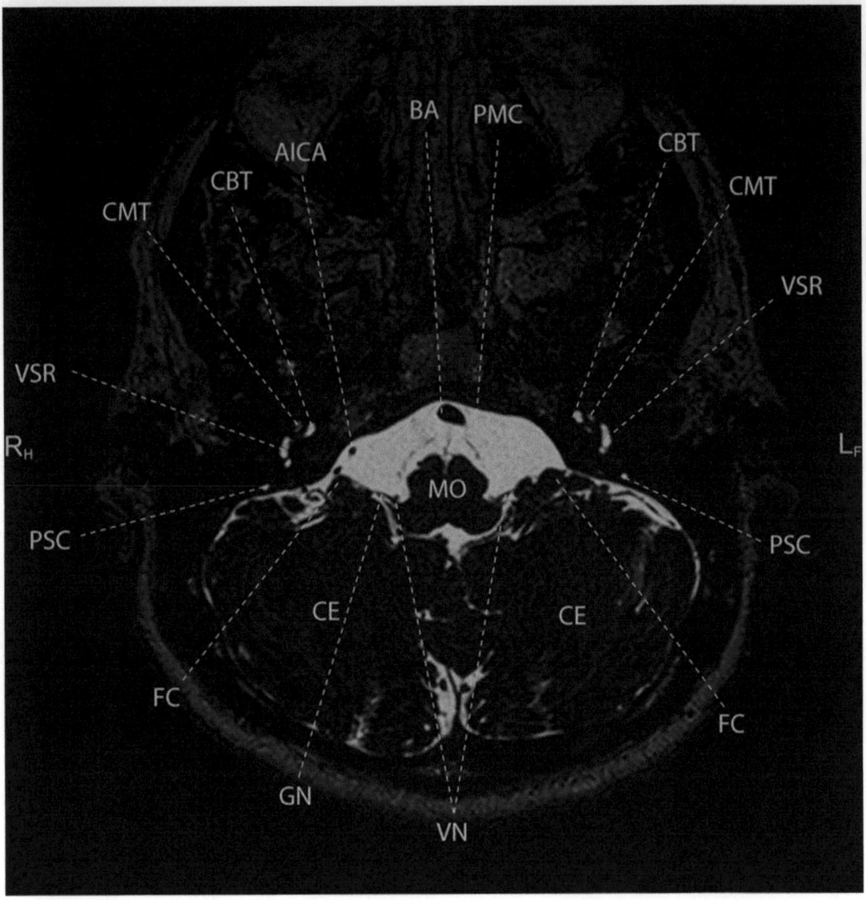

M23: MRI axial sections

AICA	anterior inferior cerebellar artery	GN	glossopharyngeal nerve
BA	basilar artery	MO	medulla oblongata
CBT	cochlea – basilar turn	PMC	premedullary cistern
CE	cerebellum	PSC	posterior semicircular canal
CMT	cochlea – middle turn	VN	vagus nerve
FC	flocculus	VSR	vestibule – spherical recess

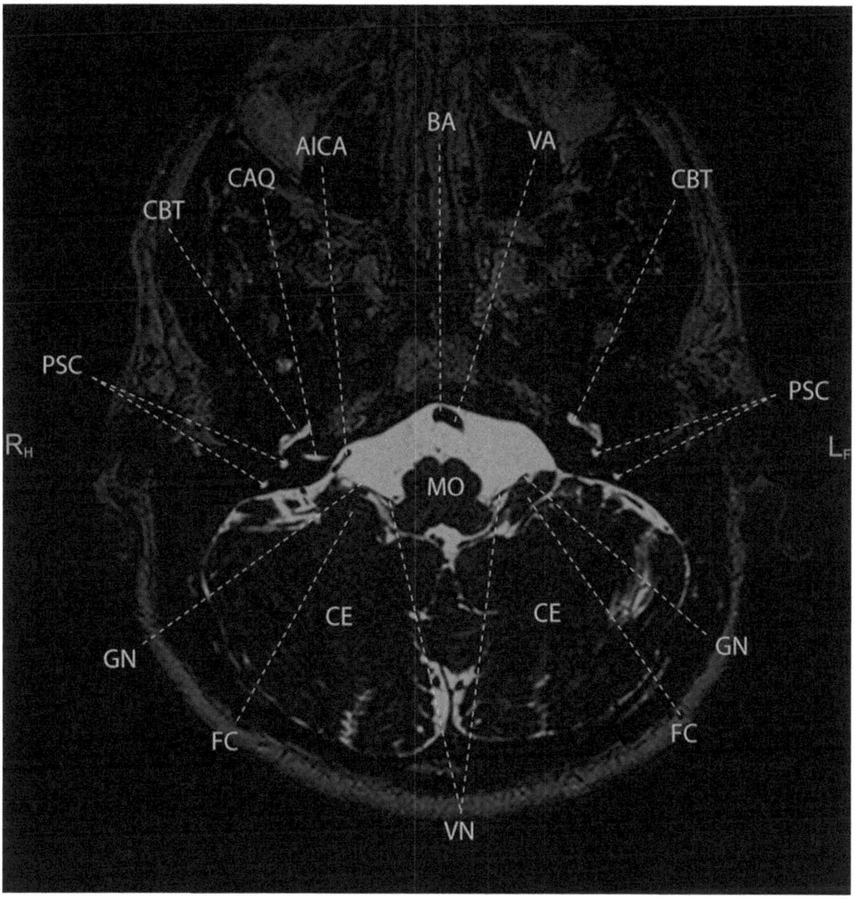

M24: MRI axial sections

AICA	anterior inferior cerebellar artery	GN	glossopharyngeal nerve
BA	basilar artery	MO	medulla oblongata
CAQ	cochlear aqueduct	PSC	posterior semicircular canal
CBT	cochlea – basilar turn	VA	vertebral artery
CE	cerebellum	VN	vagus nerve
FC	flocculus		

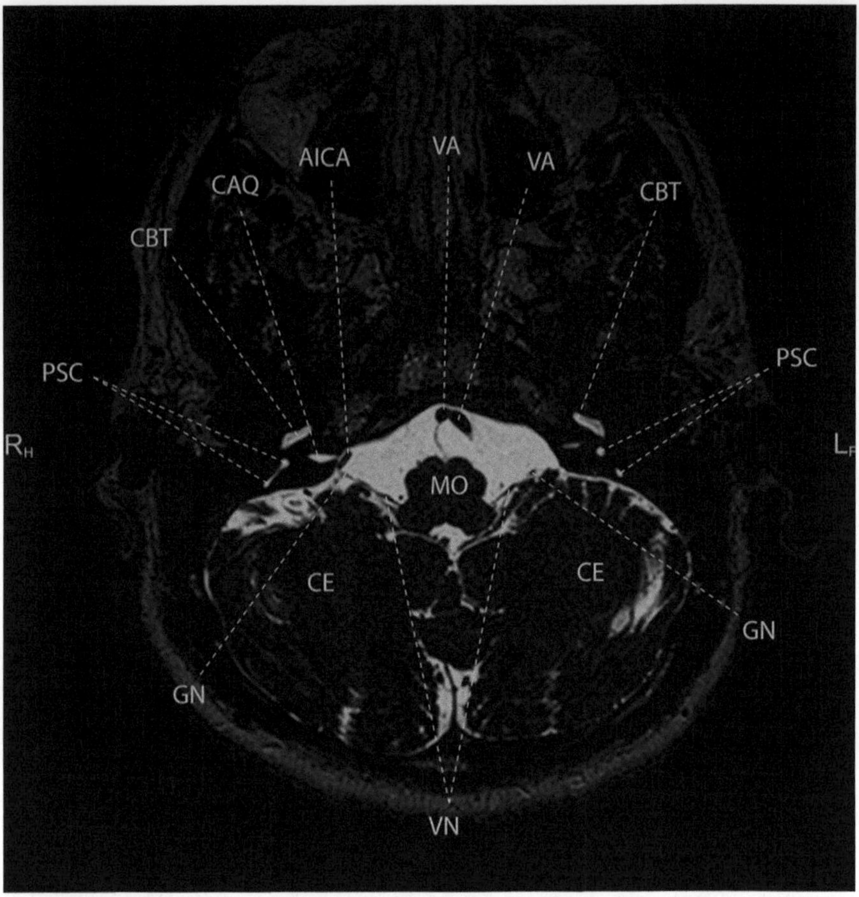

M25: MRI axial sections

AICA	anterior inferior cerebellar artery	MO	medulla oblongata
CAQ	cochlear aqueduct	PSC	posterior semicircular canal
CBT	cochlea – basilar turn	VA	vertebral artery
CE	cerebellum	VN	vagus nerve
GN	glossopharyngeal nerve		

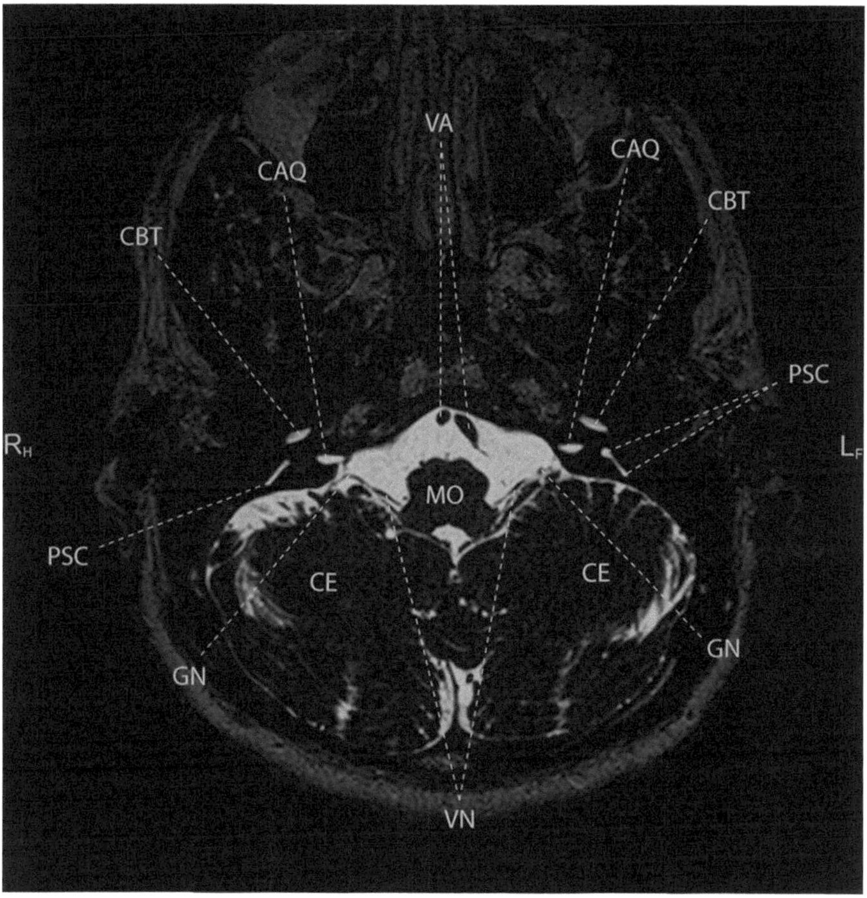

M26: MRI axial sections

CAQ	cochlear aqueduct	MO	medulla oblongata
CBT	cochlea – basilar turn	PSC	posterior semicircular canal
CE	cerebellum	VA	vertebral artery
GN	glossopharyngeal nerve	VN	vagus nerve

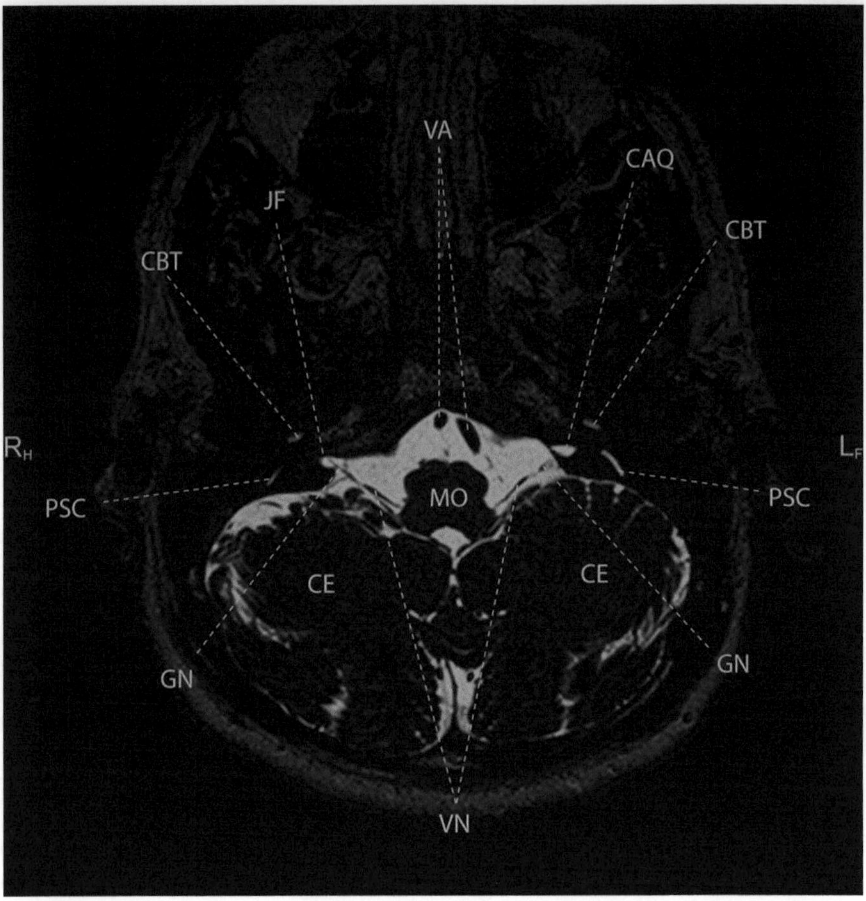

M27: MRI axial sections

CAQ	cochlear aqueduct	MO	medulla oblongata
CBT	cochlea – basilar turn	PSC	posterior semicircular canal
CE	cerebellum	VA	vertebral artery
GN	glossopharyngeal nerve	VN	vagus nerve
JF	jugular foramen		

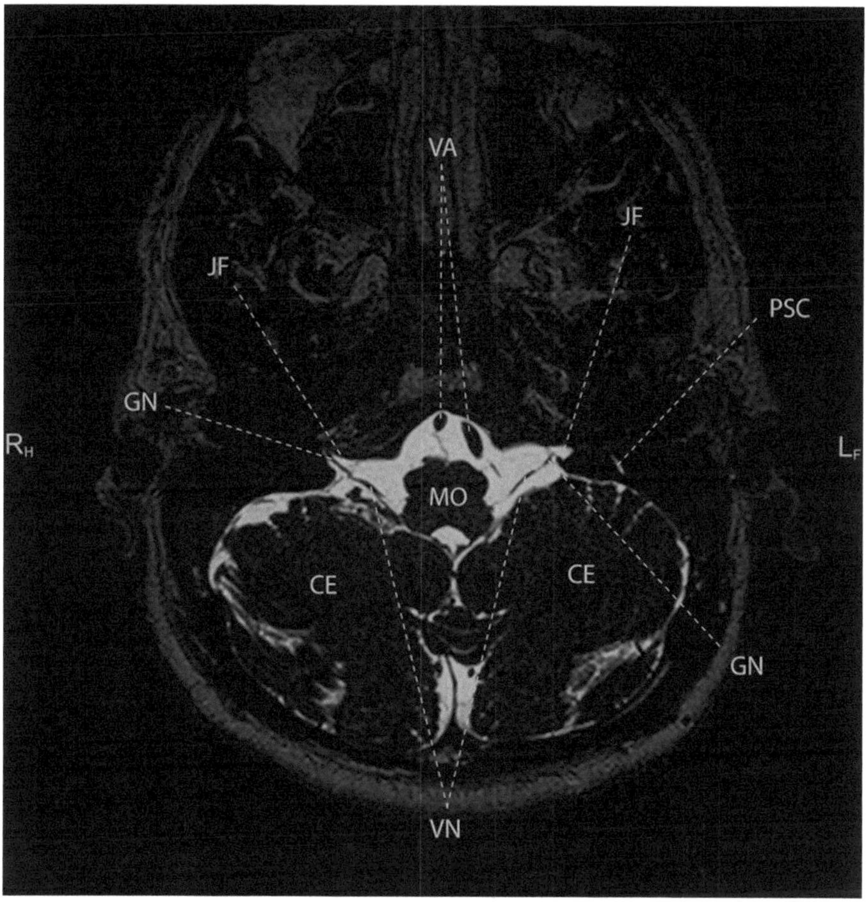

M28: MRI axial sections

CE	cerebellum	PSC	posterior semicircular canal
GN	glossopharyngeal nerve	VA	vertebral artery
JF	jugular foramen	VN	vagus nerve
MO	medulla oblongata		

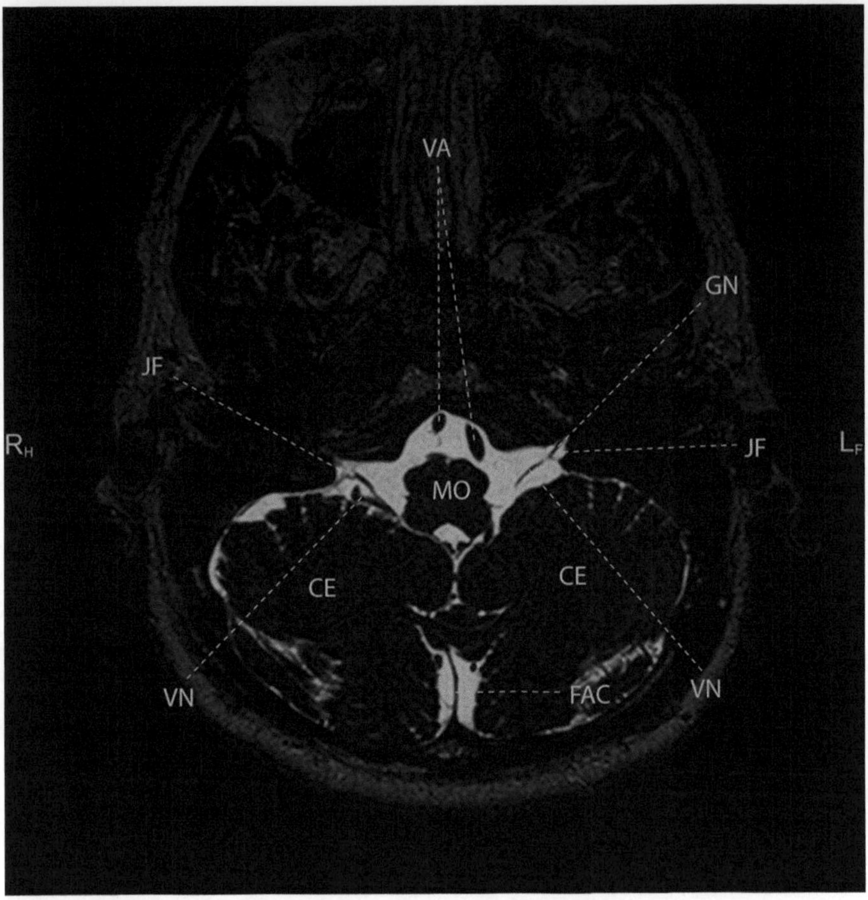

M29: MRI axial sections

CE	cerebellum	MO	medulla oblongata
FAC	falx cerebelli	VA	vertebral artery
GN	glossopharyngeal nerve	VN	vagus nerve
JF	jugular foramen		

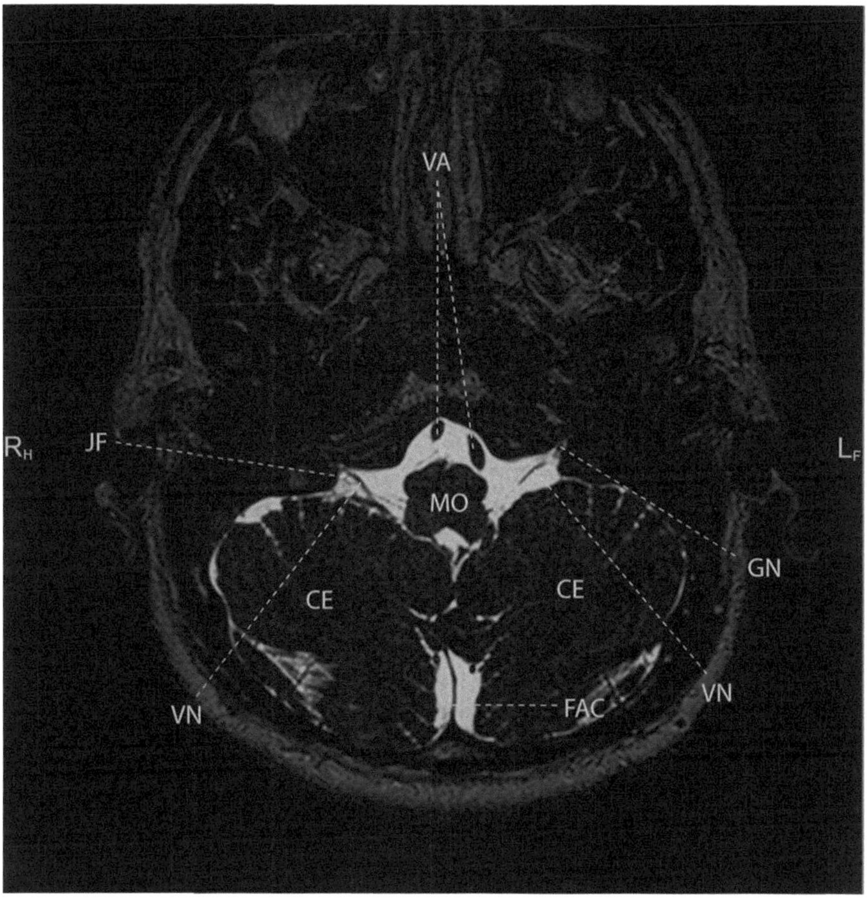

M30: MRI axial sections

CE	cerebellum	MO	medulla oblongata
FAC	falx cerebelli	VA	vertebral artery
GN	glossopharyngeal nerve	VN	vagus nerve
JF	jugular foramen		

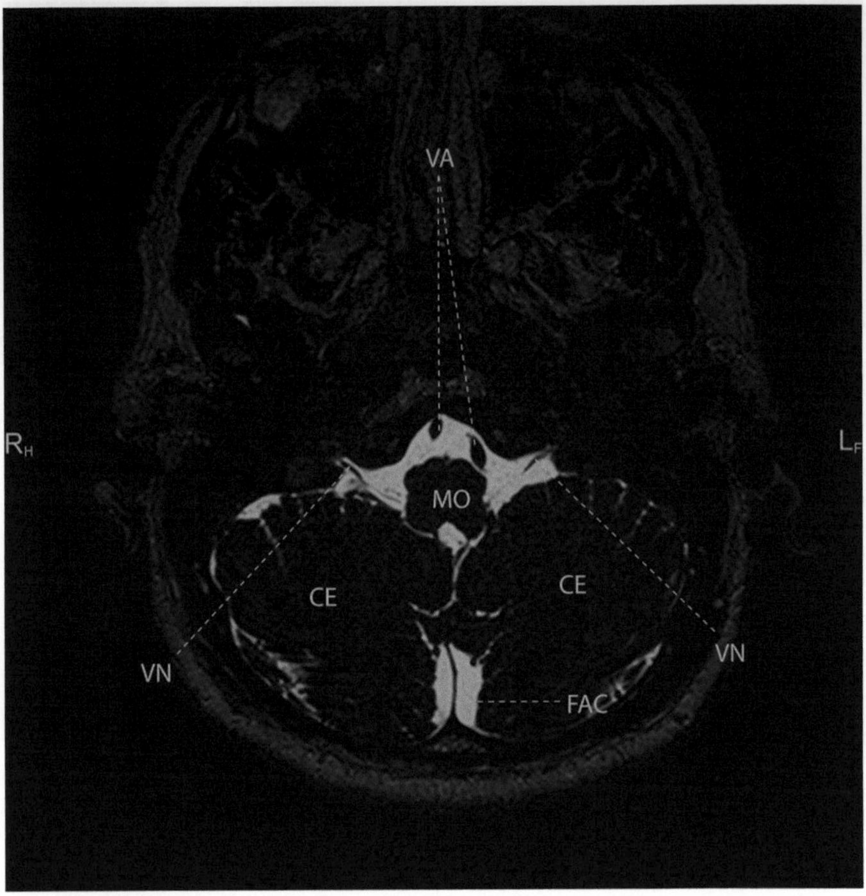

M31: MRI axial sections

CE	cerebellum	VA	vertebral artery
FAC	falx cerebelli	VN	vagus nerve
MO	medulla oblongata		

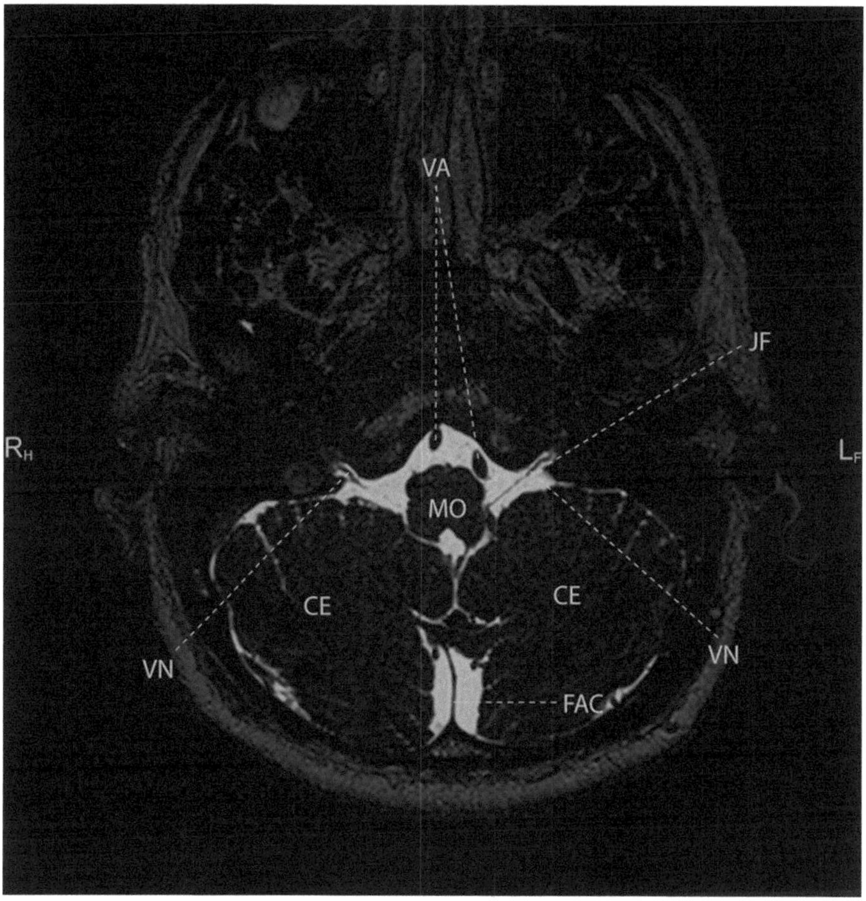

M32: MRI axial sections

CE	cerebellum	MO	medulla oblongata
FAC	falx cerebelli	VA	vertebral artery
JF	jugular foramen	VN	vagus nerve

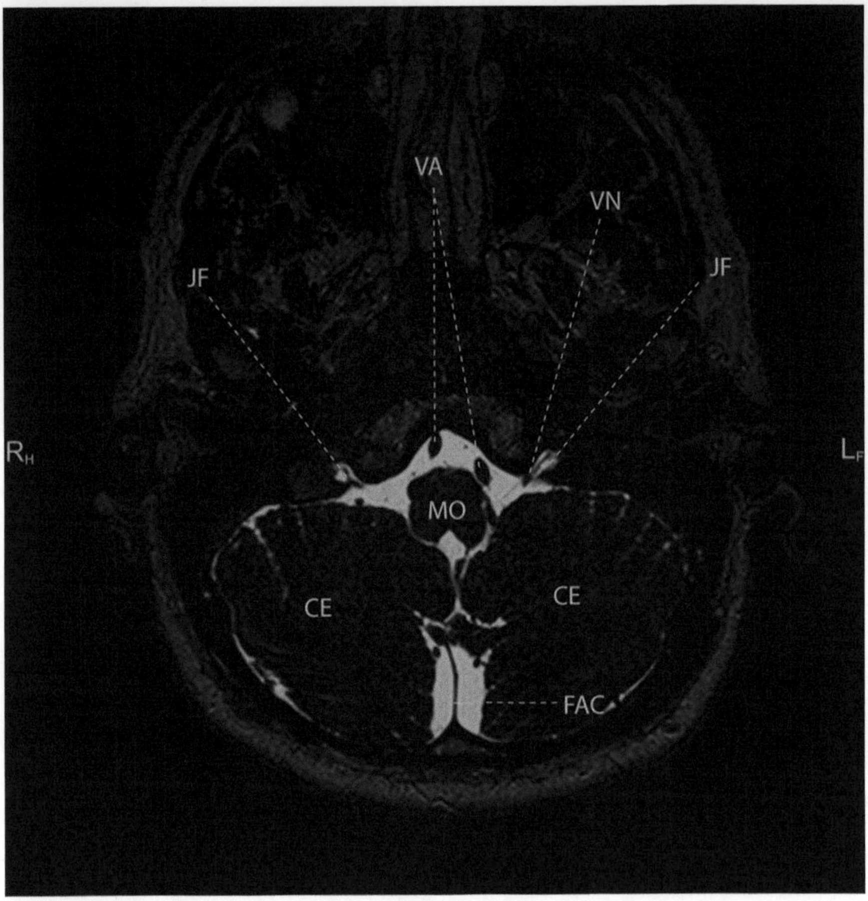

M33: MRI axial sections

CE	cerebellum	MO	medulla oblongata
FAC	falx cerebelli	VA	vertebral artery
JF	jugular foramen	VN	vagus nerve

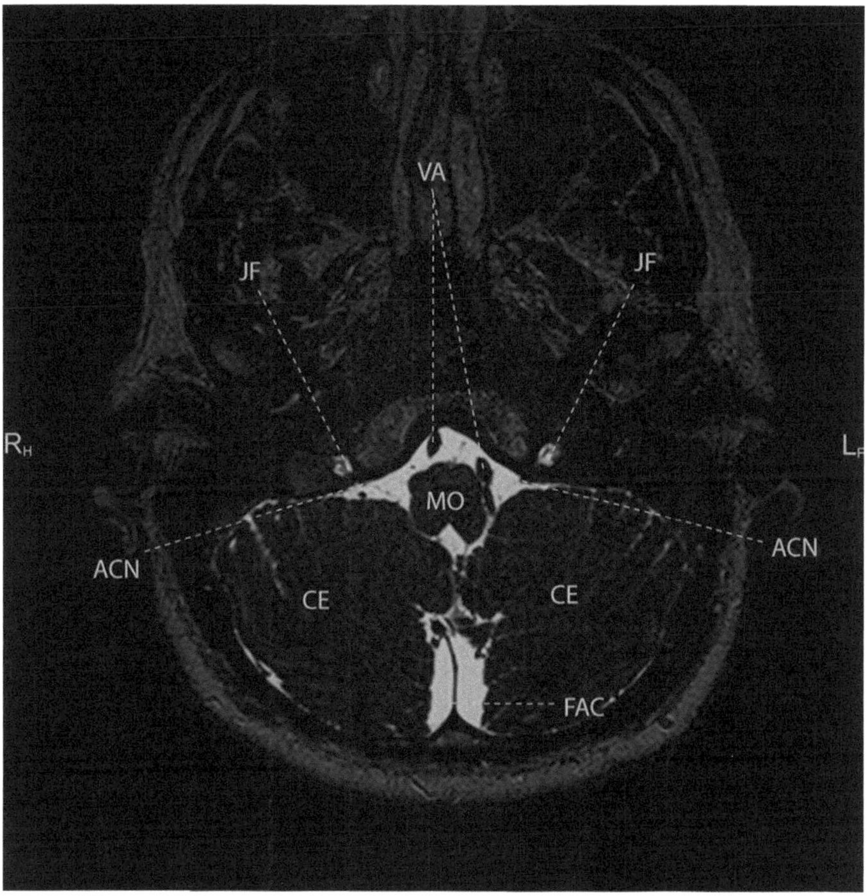

M34: MRI axial sections

ACN	accessory nerve	JF	jugular foramen
CE	cerebellum	MO	medulla oblongata
FAC	falx cerebelli	VA	vertebral artery

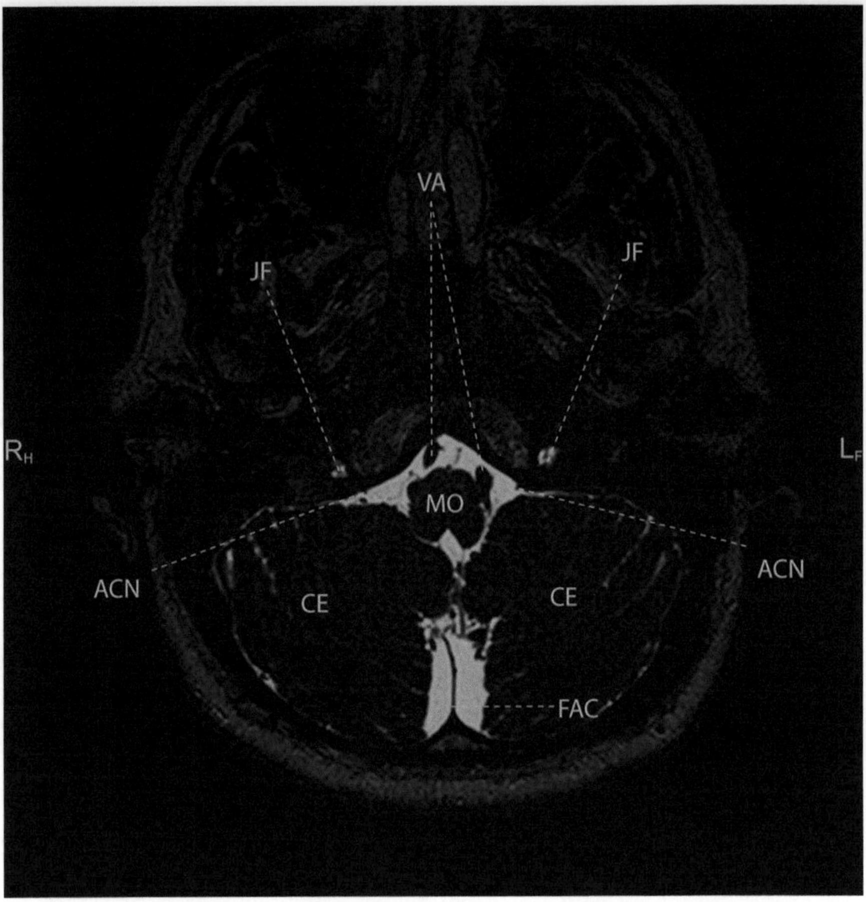

M35: MRI axial sections

ACN	accessory nerve	JF	jugular foramen
CE	cerebellum	MO	medulla oblongata
FAC	falx cerebelli	VA	vertebral artery

Suggested Reading

Abrahams PH, Spratt JD, Boon J (2008) McMinn's clinical atlas of human anatomy. Mosby Elsevier, Philadelphia

Chrobok V, Pellant A, Profant M (eds) (2008) Cholesteatoma of the temporal bone. Tobiáš, Havlíčkův Brod

Dauber W, Spitzer G, Kramer S (2007) Pocket atlas of human anatomy. Founded by Heinz Feneis. Thieme, Stuttgart/New York

International Commission on Radiological Protection (2011) Statement on tissue reactions. International Commission on Radiological Protection, ref 4825-3093-1464, Seoul. Available at https://rpop.iaea.org/RPOP/RPoP/Content/Documents/Whitepapers/ICRP-statements-tissue-reactions.pdf

Kopf-Maier P, Wolf-Heidegger G (eds) (2005) Wolf-Heidegger's atlas of human anatomy. Karger, Basel

Lane JI, Witte RJ (2010) The temporal bone. An imaging atlas. Springer, Berlin/Heidelberg/New York

Mafee MF, Valvassori GE, Becker M (2004) Imaging of the head and neck. Thieme, Stuttgart/New York

Mettler FA Jr, Bhargavan M, Faulkner K et al (2009) Radiologic and nuclear medicine studies in the United States and worldwide: frequency, radiation dose, and comparison with other radiation sources 1950–2007. Radiology 253:520–531

Putz R, Pabst R (2008) Sobotta atlas of human anatomy. Elsevier, Munich

Ryska P, Kvasnicka T, Jandura J, Klzo L, Grepl J, Zizka J (2014) Reduction of effective dose and organ dose to the eye lens in head MDCT using iterative image reconstruction and automatic tube current modulation. Biomed Pap Med Fac Univ Palacky Olomouc Czech Repub 158(2):265–272. doi:10.5507/bp.2013.071

Smith-Bindman R, Lipson J, Marcus R et al (2009) Radiation dose associated with common computed tomography examinations and the associated lifetime attributable risk of cancer. Arch Intern Med 169:2078–2086

Swartz JD, Loevner LA (2008) Imaging of the temporal bone. Thieme, Stuttgart/New York

Truwit CL, Lempert TE (1994) High resolution atlas of cranial neuroanatomy. Williams & Wilkins, Baltimore

J. Kopřiva, J. Žižka, *Temporal Bone CT and MRI Anatomy: A Guide to 3D Volumetric Acquisitions*, DOI 10.1007/978-3-319-08242-4

Index

J. Kopřiva, J. Žižka, *Temporal Bone CT and MRI Anatomy: A Guide to 3D Volumetric Acquisitions*, DOI 10.1007/978-3-319-08242-4

MIX
Papier aus verantwortungsvollen Quellen
Paper from responsible sources
FSC® C105338

If you have any concerns about our products,
you can contact us on
ProductSafety@springernature.com

In case Publisher is established outside the EU,
the EU authorized representative is:
Springer Nature Customer Service Center GmbH
Europaplatz 3, 69115 Heidelberg, Germany

Printed by Libri Plureos GmbH
in Hamburg, Germany